MONOGRAPHS ON STATISTICS AND APPLIED PROBABILITY

General Editors

D.R. Cox, D.V. Hinkley, D. Rubin and B.W. Silverman

1 Stochastic Population Models in Ecology and Epidemiology *M.S. Bartlett* (1960)

2 Queues *D.R. Cox and W.L. Smith* (1961)

3 Monte Carlo Methods *J.M. Hammersley and D.C. Handscomb* (1964)

4 The Statistical Analysis of Series of Events *D.R. Cox and P.A.W. Lewis* (1966)

5 Population Genetics *W.J. Ewens* (1969)

6 Probability, Statistics and Time *M.S. Bartlett* (1975)

7 Statistical Inference *S.D. Silvey* (1975)

8 The Analysis of Contingency Tables *B.S. Everitt* (1977)

9 Multivariate Analysis in Behavioural Research *A.E. Maxwell* (1977)

10 Stochastic Abundance Models *S. Engen* (1978)

11 Some Basic Theory for Statistical Inference *E.J.G. Pitman* (1979)

12 Point Processes *D.R. Cox and V. Isham* (1980)

13 Identification of Outliers *D.M. Hawkins* (1980)

14 Optimal Design *S.D. Silvey* (1980)

15 Finite Mixture Distributions *B.S. Everitt and D.J. Hand* (1981)

16 Classification *A.D. Gordon* (1981)

17 Distribution-free Statistical Methods *J.S. Maritz* (1981)

18 Residuals and Influence in Regression *R.D. Cook and S. Weisberg* (1982)

19 Applications of Queueing Theory *G.F. Newell* (1982)

20 Risk Theory, 3rd edition *R.E. Beard, T. Pentikäinen and E. Pesonen* (1984)

21 Analysis of Survival Data *D.R. Cox and D. Oakes* (1984)

22 An Introduction to Latent Variable Models *B.S. Everitt* (1984)

23 Bandit Problems *D.A. Berry and B. Fristedt* (1985)

24 Stochastic Modelling and Control *M.H.A. Davis and R. Vinter* (1985)

25 The Statistical Analysis of Compositional Data *J. Aitchison* (1986)

26 Density Estimation for Statistics and Data Analysis *B.W. Silverman* (1986)

27 Regression Analysis with Applications *G.B. Wetherill* (1986)

28 Sequential Methods in Statistics, 3rd edition *G.B. Wetherill* (1986)

29 Tensor Methods in Statistics *P. McCullagh* (1987)

30 Transformation and Weighting in Regression *J.R. Carroll and D. Ruppert* (1988)

31 Asymptotic Techniques for Use in Statistics *O.E. Barndorff-Nielsen and D.R. Cox* (1989)

32 Analysis of Binary Data, 2nd edition *D.R. Cox and E.J. Snell* (1989)

33 Analysis of Infectious Disease Data *N.G. Becker* (1989)

34 Design and Analysis of Cross-Over Trials *B. Jones and M.G. Kenward* (1989)

35 Empirical Bayes Methods, 2nd edition *J.S. Maritz and T. Lwin* (1989)

36 Symmetric Multivariate and Related Distributions *K.-T. Fang, S. Kotz and K.W. Ng* (1989)

37 Generalized Linear Models, 2nd edition *P. McCullagh and J.A. Nelder* (1989)

38 Cyclic Designs *J.A. John* (1987)

39 Analog Estimation Methods in Econometrics *C.F. Manski* (1988)

(Full details concerning this series are available from the Publishers)

Symmetric Multivariate and Related Distributions

KAI-TAI FANG
Institute of Applied Mathematics, Academia Sinica, China

SAMUEL KOTZ
Department of Management Science and Statistics, University of Maryland

KAI WANG NG
Department of Statistics, University of Hong Kong

LONDON NEW YORK
CHAPMAN AND HALL

First published in 1990 by
Chapman and Hall Ltd
11 New Fetter Lane, London EC4P 4EE
Published in the USA by
Chapman and Hall
29 West 35th Street, New York NY 10001

Typeset in 10/12pt Times by Thomson Press (India) Ltd, New Delhi
Printed in Great Britain by St. Edmundsbury Press Ltd,
Bury St. Edmunds, Suffolk

ISBN 0 412 314304

British Library Cataloguing in Publication Data

Fang, Kai-Tang
Symmetric multivariate and related distributions.
1. Multivariate analysis
I. Title. II. Kotz, Samuel. III. Ng, Kai W. IV. Series.
519.5′35

ISBN 0-412-31430-4

Library of Congress Cataloging in Publication Data

Fang, Kai-Tang.
Symmetric multivariate and related distributions / Kai-Tang Fang, Samuel Kotz, Kai W. Ng.
p. cm. -- (Monographs on statistics and applied probability)
Bibliography: p.
Includes index.
ISBN 0-412-31430-4
1. Distribution (Probability theory) I. Kotz, Samuel. II. Ng, Kai W. III. Title. IV. Series.
QA273.6.F36 1989
519.2′4—dc20 89-32953
CIP

To our wives: Tingmei, Rosalie and May for their love, constant support, and understanding that sometimes symmetric multivariate distributions may come first.

Contents

Preface

This book represents the joint effort of three authors, located thousands of miles apart from one another, and its preparation was greatly facilitated by modern communication technology. It may serve as an additional example of international co-operation among statisticians specializing in statistical distribution theory.

In essence we have attempted to amass and digest widely scattered information on multivariate symmetric distributions which has appeared in the literature during the last two decades. Introductory remarks at the beginning of each chapter summarize its content and clarify the importance and applicability of the distributions discussed in these chapters; it seems unnecessary, therefore, to dwell in this Preface on the content of the volume.

It should be noted that this work was initiated by the first author, who provided continuous impetus to the project, and a great many of the results presented in the book stem from his own research or the research of his associates and students during the last 15 years. Some of the original contributions of the second author, who also guided the literature search, in the field of multivariate distributions derived a decade ago have been included in substantially extended and generalized form. Contributions of the third author are equally important in providing a coherent description of novel results discussed in the last chapters of this volume, which incorporate some of his recent results on the topics, as well as in co-ordinating the whole project and shaping it into the typescript form. The authors enjoyed working as a team and each one contributed to all the chapters albeit in varying degrees.

Compilation of this work was preceded by painstaking and comprehensive literature search and study. Our most sincere thanks are to anonymous librarians in many parts of the USA, England, Europe, Peoples' Republic of China, Hong Kong and Canada, who assisted us, far beyond the call of duty. Our thanks are also due to numerous authors (far too numerous to cite individually) who generously supplied us with preprints and reprints of their recent works in the area. Our special thanks are due to Sir David R. Cox, who brought our work to the attention of Chapman and Hall Publishing House and whose pioneering contributions in numerous branches of statistics and distribution theory significantly inspired our undertaking. We are thankful to Ms Elizabeth Johnston, Senior Editor at Chapman and Hall, for her capable, experienced and patient guidance related to editorial and technical matters and for facilitating communication among the authors. We are grateful to the Chairman of the Department of Statistics at the University of North Carolina, Professor S. Cambanis (an important contributor to the field of multivariate symmetric distributions), and Professor P.M. Bentler, Department of Psychology, University of California at Los Angeles, for the hospitality afforded the first author, and to York University, Canada, for providing the third author with congenial facilities during his one-year visit at the Department of Mathematics. Last but not least, we are happy to acknowledge the skilful and efficient word-processing and typing service of Mrs Lucile Lo and Miss Irene Cheung.

The authors sincerely hope that their endeavour will stimulate additional research in the field of multivariate symmetric distributions and will contribute to the rapid penetration of these distributions into the field of applied multivariate analysis as a more versatile alternative to the classical methods based on multivariate normal distributions.

Regrettably due to constraints on the size of this book, problems dealing with sampling from symmetric elliptical and related distributions are not discussed in this volume. The authors hope to tackle these important topics in a future publication.

Kai-Tai Fang
Institute of Applied Mathematics, Academia Sinica
P.O. Box 2734
100080 Beijing, China.

Samuel Kotz
Department of Management Science and Statistics
University of Maryland
College Park, Md. 20742, USA.

Kai W. Ng
Department of Statistics
University of Hong Kong, Hong Kong.

CHAPTER 1

Preliminaries

In this chapter, a unified approach to the construction of symmetric multivariate distributions is presented. This approach provides the reader with clear guidelines for developing the theory of symmetric multivariate distributions. This first section deals with these matters. Sections 1.2 and 1.3 deal with notation and some of the mathematical tools required in what follows. Finally, in Section 1.4 we present a fairly complete discussion of the Dirichlet distribution which is an important multivariate distribution on its own as well as a useful tool for the development of other classes of multivariate distributions.

Readers familiar with these topics may find the pace in this chapter a bit too leisurely and may wish to skim through it. Our experience shows, however, that it is important to lay a sound mathematical foundation carefully as well as intuitive motivation at the beginning in order to facilitate the understanding of more subtle and involved arguments which will appear in succeeding chapters. For a more comprehensive and advanced treatment of many of the notions discussed in this chapter, the reader is referred to the monograph by Fang and Zhang, *Generalized Multivariate Analysis* (1989).

1.1 Construction of symmetric multivariate distributions

Since the turn of the twentieth century, statisticians have been engaged in extending univariate distributions to their multivariate analogues. There are numerous approaches to this problem. In the survey paper, Multivariate distributions at a cross-road, written by Kotz (1975) and in the joint paper of Goodman and Kotz (1981), Samuel Kotz presented a more or less systematic classification of these methods of extension of multivariate distributions based on various criteria such as the type of dependence, analogy of mathematical form, model and characterizations. Although these problems have not yet been completely and fully solved, substantial progress has been made in this area. Researchers in statistical distributions have succeeded in developing, in the last three or four decades, numerous univariate distributions, shifting the centre of gravity of univariate distributions from the normal distribution. Encouraged by this success, investigators have more recently turned their attention and methodology to an analysis of multivariate distributions other than the normal one.

As one would suspect, the 'jungle' of non-normality in the multivariate setting is not easily tamed. Nevertheless, substantial progress has been made on several fronts and part of our purpose is to survey some of them as well as to contribute additional elucidation. We shall confine ourselves to a more modest but important task of multivariate symmetric distributions. The multivariate aspect characterizes the model of dependent outcomes of random experiments while symmetry (and the closely related concept of invariance which will also appear prominently in our deliberations – see Section 1.3) goes, of course, to the heart of the theory and philosophy of the laws of nature and mathematics. (The interested reader is referred to the well-known fascinating monograph by Hermann Weyl, *Symmetry* (1952) and to Wigner (1967).) We also note that it was B. Efron (1969) who pointed out that the weaker condition of symmetry can successfully replace the normal distribution as far as the classical Student's t-test is concerned, although as early as 1925 R.A. Fisher showed that the rotational symmetry (see Section 1.3) of the random vector

$$\mathbf{x} = (x_1, x_2, \ldots, x_n)$$

under the null hypothesis $\mu = 0$ is all that is necessary to yield the standard null distribution of Student's one-sample t-statistics.

For definiteness assume that we wish to extend the normal distribution $N(\mu, \sigma^2)$ to the multivariate case. The following are some possible avenues.

1. *By means of density.*
Since the density of $x \sim N(\mu, \sigma^2)$ is

$$\frac{1}{\sqrt{2\pi}\sigma}\exp\left(-\tfrac{1}{2}(x-\mu)^2/\sigma^2\right) = \frac{1}{\sqrt{2\pi}\sigma}\exp\left(-\tfrac{1}{2}(x-\mu)(\sigma^2)^{-1}(x-\mu)\right),$$

we can guess that the density of multivariate normal distributions should be of the form

$$C|\mathbf{\Sigma}|^{-1/2}\exp\{-\tfrac{1}{2}(\mathbf{x}-\boldsymbol{\mu})'\mathbf{\Sigma}^{-1}(\mathbf{x}-\boldsymbol{\mu})\}, \tag{1.1}$$

where $\mathbf{x}$, $\boldsymbol{\mu}\in R^n$, $\mathbf{\Sigma}$:$n\times n$, $\mathbf{\Sigma}>0$ and C is a normalizing constant.

2. *By means of the characteristic function* (*c.f.*).
It is well known that the c.f. of $N(\mu,\sigma^2)$ is

$$\exp\{it\mu-\tfrac{1}{2}\sigma^2t^2\}=\exp\{it\mu-\tfrac{1}{2}t(\sigma^2)t\}.$$

Thus an appropriate form of the c.f. of the multivariate normal distribution can be

$$\exp\{i\mathbf{t}'\boldsymbol{\mu}-\tfrac{1}{2}\mathbf{t}'\mathbf{\Sigma}\mathbf{t}\}, \tag{1.2}$$

where $\mathbf{t}$, $\boldsymbol{\mu}\in R^n$ and $\mathbf{\Sigma}>0$.

3. *By a linear combination.*
Let $\mathbf{x}$ be an $n\times 1$ random vector. If for each $\mathbf{a}\in R^n$, $\mathbf{x}'\mathbf{a}$ is normal, we say that $\mathbf{x}$ possesses a multivariate normal distribution. Obviously this last method can be successful only for a few selective distributions.

4. *By means of stochastic decomposition.*
If $x\sim N(\mu,\sigma^2)$, it can be expressed as

$$x=\mu+\sigma y,$$

where $y\sim N(0,1)$. Thus, if we can extend the standard normal distribution to the multivariate case, we can also obtain the general multivariate normal distribution. Let $\mathbf{y}=(y_1,\ldots,y_m)'$ have independent identically distributed (i.i.d.) components and $y_1\sim N(0,1)$. The distribution of $\mathbf{y}$ is a natural multivariate extension of the standard normal distribution. Let

$$\mathbf{x}=\boldsymbol{\mu}+\mathbf{A}\mathbf{y}, \tag{1.3}$$

where $\mathbf{x},\boldsymbol{\mu}\in R^n$ and $\mathbf{A}$:$n\times m$. We postulate that $\mathbf{x}$ is distributed according to a multivariate normal distribution, and write $\mathbf{x}\sim N_n(\boldsymbol{\mu},\mathbf{\Sigma})$ where $\mathbf{\Sigma}=\mathbf{A}\mathbf{A}'$.

There are numerous indications that the last approach is the most convenient and fruitful. In the early stages, investigators tried to extend univariate distributions to a multivariate setting using the so-called 'one by one' method. A comprehensive survey of multivariate continuous distributions is given in Johnson and Kotz's book (1972), in which detailed discussions of many specific multivariate distributions are presented. A careful reader will note that most of them, in the standard form when stripped of their parameters (such as $\boldsymbol{\mu}$, $\boldsymbol{\Sigma}$, etc.), are symmetric in a certain sense.

More recently, symmetric multivariate distributions have appeared at an ever increasing rate. This is particularly true in the study of distribution theory, generalizations of sampling distributions and statistics, and the theory of robust estimation (as stimulated by the above-mentioned Efron's paper), Bayesian inference, projection pursuit and so on. In fact, many inference problems have the naturally inherent properties of symmetry.

A symmetric multivariate version of the univariate distribution may be obtained not only as a mathematical extension, but also from natural statistical considerations.

First consider a possible extension of the classical sampling theory. Let $x_1, \ldots, x_n$ be a sample of $N(0, \sigma^2)$, $\sigma > 0$, and

$$t_x = \sqrt{n}\frac{\bar{x}}{\sigma_x}, \tag{1.4}$$

where

$$\bar{x} = \frac{1}{n}\sum_1^n x_i$$

$$s_x^2 = \frac{1}{n-1}\sum_1^n (x_i - \bar{x})^2.$$

It is clear that $t_x \sim t_{n-1}$, the t-distribution with $(n-1)$ degrees of freedom. Let $r > 0$ be a real-valued random variable and

$$\mathbf{y} = r\mathbf{x} = \begin{bmatrix} rx_1 \\ \vdots \\ rx_n \end{bmatrix} = \begin{bmatrix} y_1 \\ \vdots \\ y_n \end{bmatrix}.$$

Similarly, we can define

$$t_y = \sqrt{n}\frac{\bar{y}}{s_y}, \tag{1.5}$$

with

$$\bar{y} = \frac{1}{n}\sum_1^n y_i = r\bar{x}$$

$$s_y^2 = \frac{1}{n-1}\sum_1^n (y_i - \bar{y})^2 = r^2 s_x^2.$$

Therefore,

$$t_y = \sqrt{n}\frac{\bar{y}}{s_y} = \sqrt{n}\frac{r\bar{x}}{rs_x} = \sqrt{n}\frac{\bar{x}}{s_x} = t_X$$

has the same distribution as t_x. Now, however, $y_1, \ldots, y_n$ constitute a set of variables from a nonnormal population. If we consider $(y_1, \ldots, y_n)$ to be an independent sample or a sample of some other type, we can then develop not only a sampling theory, but also derive a symmetric multivariate distribution, that of $\mathbf{y} = (y_1, \ldots, y_n)'$.

Let us now explore the meaning of a symmetry of a multivariate distribution. It is evident from the classical sampling theory that the first prerequisite for symmetry is exchangeability of the components. We say $\mathbf{y} = (y_1, \ldots, y_n)$ is **exchangeable** if

$$(y_{i_1}, \ldots, y_{i_n}) \stackrel{d}{=} (y_1, \ldots, y_n) \tag{1.6}$$

for each permutation $(i_1, \ldots, i_n)$ of $(1, \ldots, n)$. Here $\stackrel{d}{=}$ signifies that the two sides have the same distribution (cf. Section 1.2). However, exchangeability alone is too broad and is not sufficient for $\mathbf{x}$ to possess adequate and natural properties. A stronger version of symmetry may be needed when the symmetry is imposed on the mathematical representation of the underlying distribution. The following are some additional types of symmetry which may be applied for constructing symmetric multivariate distributions.

1. *Symmetry guided by invariance of the distribution under a group of transformations.*

Let $\mathbf{x}$ be an n-dimensional random vector and $\mathcal{O}(n)$ be the set of $n \times n$ orthogonal matrices. $\mathcal{O}(n)$ is a group (see Section 1.3 for more details), called the **orthogonal group**; here the group operation is matrix multiplication.

If it is true that

$$\mathbf{Hx} \stackrel{d}{=} \mathbf{x} \quad \text{for each} \quad \mathbf{H} \in \mathcal{O}(n), \tag{1.7}$$

we shall say that $\mathbf{x}$ possesses a **spherical distribution**. Below we shall see that a $\mathbf{x} \sim N_n(\mathbf{0}, \sigma^2\mathbf{I}_n)$ has a spherical distribution (cf. Chapters 2, 3 and 4).

Let $\mathscr{P}$ be the **permutation group**, i.e. if $\mathbf{H} \in \mathscr{P}$, then $\mathbf{H}'\mathbf{H} = \mathbf{I}$ and the elements of $\mathbf{H}$ are only 0 or 1. The exchangeability as defined by

(1.6) is equivalent to the condition

$$\mathbf{Hx} \overset{d}{=} \mathbf{x} \quad \text{for each} \quad \mathbf{H} \in \mathscr{P}.$$

Evidently the group $\mathscr{P}$ is a subgroup of the orthogonal group of the corresponding dimension.

2. *Symmetry of the cumulative distribution function.*
A multivariate function $F(x_1, \ldots, x_n)$ is said to be **symmetric** if for each permutation $(i_1, \ldots, i_n)$ of $(1, \ldots, n)$ we have

$$F(x_{i_1}, \ldots, x_{i_n}) = F(x_1, \ldots, x_n).$$

When F is the cumulative distribution function (c.d.f.) of a random vector $\mathbf{x}$, the symmetry of F is equivalent to (1.6), hence $\mathbf{x}$ is exchangeable. In particular, if

$$F(x_1, \ldots, x_n) = \prod_{i=1}^{n} G(x_i),$$

then $\mathbf{x}$ has i.i.d. components with c.d.f. $G(\cdot)$, and is exchangeable. The 'i.i.d.' is a classical symmetry.

3. *Symmetry of the characteristic function* (*c.f.*).
The c.f. of a stable law of index α is in the form of

$$\exp(-c|t|^{\alpha}), \quad 0 < \alpha \leqslant 2, \quad c > 0.$$

For $\alpha = 1$, it is the c.f. of a Cauchy distribution, and for $\alpha = 2$, it becomes the c.f. of the standard normal distribution. If $(x_1, \ldots, x_n)$ is a sample from a stable law, its c.f. is given by

$$\exp\left(-c\sum_{1}^{n} |t_j|^{\alpha}\right). \tag{1.8}$$

A natural generalization of (1.8) is to a c.f. of the form

$$\phi(|t_1|^{\alpha} + \cdots + |t_n|^{\alpha}). \tag{1.9}$$

If the c.f. of $\mathbf{x}$ is of the form (1.9) we say that $\mathbf{x}$ possesses an **α-symmetric multivariate distribution** (cf. Chapter 7).

For $\alpha = 2$, the 2-symmetric multivariate distribution reduces to the spherical distribution.

4. *Symmetry of the survival function.*
It is well known that the survival function of the exponential distribution with parameter λ (to be denoted $E(\lambda)$) with density $\lambda e^{-\lambda x}$ is

$$P(x > u) = e^{-\lambda u}, \quad u > 0; \quad \lambda > 0.$$

If $\mathbf{x} = (x_1, \ldots, x_n)$ is a sample from $E(\lambda)$, the joint survival function of $(x_1, \ldots, x_n)$ is

$$\begin{aligned} P(\mathbf{x} > \mathbf{u}) &= P(x_1 > u_1, \ldots, x_n > u_n) = \exp\left(-\lambda \sum_1^n u_i \right) \\ &= \exp(-\lambda \| \mathbf{u} \|), \end{aligned}$$

where the norm is the ℓ_1-sense.

We could generalize this to the form

$$P(\mathbf{x} > \mathbf{u}) = h(\| \mathbf{u} \|) \tag{1.10}$$

for some function h. If the survival function of a nonnegative random vector $\mathbf{x}$ is of the form (1.10), we say that $\mathbf{x}$ has an ℓ_1-norm symmetric multivariate distribution (cf. Chapter 5).

5. *Symmetry of the density.*
Let $\mathbf{x} = (x_1, \ldots, x_n)'$ be a random sample from $N(0, \sigma^2)$. The joint density of $\mathbf{x}$ is

$$c \exp\left(-\frac{1}{2\sigma^2} \mathbf{x}'\mathbf{x} \right)$$

where the variables appear solely through $\mathbf{x}'\mathbf{x}$ and c is a constant independent of $\mathbf{x}'\mathbf{x}$. We could extend this distribution by utilizing the above property. If $\mathbf{x}$ possesses a density of the form of $f(\mathbf{x}'\mathbf{x})$, for some function f, we say that x has a spherical distribution. Compare with types 1 and 3 above, where sphericity was also defined. The equivalence of the two definitions will be established below under certain conditions. Note that in general, it is not necessary that a spherical distribution possess a density. Using the density approach we obtain only a proper subset of spherical distributions.

Similarly, if $x_1, \ldots, x_n$ is a sample from $E(\lambda)$, then their joint density is

$$\lambda^n \exp\left(-\lambda \sum_1^n x_i \right) = \lambda^n \exp(-\lambda \| \mathbf{x} \|)$$

which is a function solely of $\| \mathbf{x} \|$, where $\| \cdot \|$ is the ℓ_1-norm. Utilizing this property, we can define a new class of distributions, the so-called ℓ_1-norm symmetric multivariate distributions (Chapter 5).

The symmetry of density implies that the contours of surfaces of equal density have the same shape in a given class. This gives rise to the name of **elliptically contoured distributions** which are an extension of spherical distributions constructed by introducing some additional parameters.

6. *Symmetry of an n-dimensional version.*
Let z be a random variable. If there exists an n-dimensional random vector $\mathbf{x}$ such that

$$\text{(a)}\quad c(\mathbf{a}) > 0 \qquad \text{if } \mathbf{a} \neq \mathbf{0},$$

$$\text{(b)}\quad \mathbf{a}'\mathbf{x} \overset{d}{=} c(\mathbf{a})z \qquad \text{for any } \mathbf{a} \in R^n$$

then $\mathbf{x}$ is called an n-dimensional version of z (Eaton, 1981).

It is evident that the above definition may serve as a mechanism for extending univariate distributions to the multivariate case. Since the symmetry of $\mathbf{x}$ is required, the function $c(\mathbf{a})$ and z must be symmetric in a certain sense and z must be symmetric at the origin. For example, if we wish the n-dimensional version to be spherical, $c(\mathbf{a})$ must be of the form

$$c(\mathbf{a}) = (\mathbf{a}'\mathbf{a})^{1/2}.$$

If we wish the n-dimensional version to be α-symmetric, $c(\mathbf{a})$ should be of the form

$$c(\mathbf{a}) = \left(\sum_{i=1}^{n} |a_i|^\alpha \right)^{1/\alpha}$$

where $\mathbf{a}' = (a_1, a_2, \ldots, a_n)$ or of a more general form (cf. Chapter 7).

7. *Symmetry guided by invariance of statistics.*
We have mentioned above that the statistic t_y defined in (1.5) has the same distribution as the statistic t_x constructed from the normal population. This leads us to the definition of the class of variables:

$$F_t = \{\mathbf{x} | \mathbf{x} \text{ is exchangeable and } t_x \sim t_{n-1}\}.$$

We shall see below that if $\mathbf{x}$ possesses a spherical distribution and $P(\mathbf{x} = \mathbf{0}) = 0$, then $\mathbf{x} \in F_t$. However, the class F_t is much wider than that of the spherical distributions. The structure of the class F_t has not as yet been sufficiently thoroughly investigated.

In the same vein, various classes of symmetric multivariate distributions can be defined utilizing other statistics.

8. *Symmetry of a stochastic representation.*
Let $\mathbf{y}$ have a symmetric multivariate distribution. One can define a class of symmetric multivariate distributions generated by $\mathbf{y}$ as follows:

$$F(\mathbf{y}) = \{\mathbf{x} \mid \mathbf{x} \overset{d}{=} R\mathbf{y}, \quad R \geqslant 0 \text{ is independent of } \mathbf{y}\}.$$

If $\mathbf{y} \sim N_n(\mathbf{0}, \sigma^2 \mathbf{I}_n)$, then the class $F(\mathbf{y})$ is equivalent to the class of mixtures of normal variables. If $\mathbf{y}$ is uniformly distributed on a unit sphere in R^n, $F(\mathbf{y})$ is the class of spherical distributions. If $\mathbf{y}$ is uniformly distributed on the simplex $\{\mathbf{z} \mid \mathbf{z} \in R^n_+, \|\mathbf{z}\| = 1\}$ with the ℓ_1-norm, $F(\mathbf{y})$ is then the class of ℓ_1-norm symmetric multivariate distributions (Chapter 5).

A crucial aspect of the mechanism described above is to select $\mathbf{y}$ which will yield a wide class of distributions with appealing properties. The following approach seems to yield satisfactory results:

1. Take a sample $z_1, \ldots, z_n$ from a population with the c.d.f. $F(x)$.
2. Let $\mathbf{z} = (z_1, \ldots, z_n)'$ and construct

$$y_i = z_i / \| z \|, \quad i = 1, \ldots, n \quad \text{and} \quad \mathbf{y} = (y_1, \ldots, y_n)',$$

where the norm is based on a suitably chosen metric.

For example, if $z_1, \ldots, z_n$ are obtained from $N(0, \sigma^2)$ and the norm is defined as $\|\mathbf{z}\|^2 = \mathbf{z}'\mathbf{z}$, then $\mathbf{y}$ is uniformly distributed on the unit sphere in R^n. If, however, $z_1, \ldots, z_n$ are sampled from $E(\lambda)$ and the norm is defined as $\|\mathbf{z}\| = \sum_i^n z_i$ then $\mathbf{y}$ is uniformly distributed on the simplex $\{\mathbf{z} \in R^n, \|\mathbf{z}\| = 1\}$. (See J. Aitchison's (1986) book in this series on multivariate distributions defined on a simplex.)

The method described above can be applied to non-symmetric multivariate distributions as well. For example, let $z_1, \ldots, z_n$ be independent and $z_i \sim \mathrm{Ga}(\alpha_i, 1)$ – a gamma distribution with the density function

$$(\Gamma(\alpha_i))^{-1} y^{\alpha_i - 1} e^{-y}, \quad y > 0, \quad i = 1, \ldots, n.$$

Define $y_i = z_i / \sum_{i=1}^n z_i = z_i / \| z \|$, $i = 1, \ldots, n$. Then $\mathbf{y}$ has a (non-symmetric) **Dirichlet distribution** (cf. Section 1.4) and the corresponding $F(\mathbf{y})$ is in a class of **multivariate Liouville distributions** (cf. Chapter 6). If in the above example $z_1, \ldots, z_n$ are sampled from a symmetrized gamma distribution with parameters $\alpha_1, \ldots, \alpha_n$ respectively, the corresponding class $F(\mathbf{y})$ will be that of generalized symmetric Dirichlet distributions (Fang and Fang, 1988c).

1.2 Notation

1.2.1 Notation of algebraic entities and characteristics of random quantities

Throughout this book the following notations will be used. We recommend that our readers familiarize themselves with the notation to facilitate understanding of the material in subsequent chapters.

Vector: $\mathbf{a} = \begin{bmatrix} a_1 \\ \vdots \\ a_n \end{bmatrix}$ is an n-dimensional vector (also called n-vector) with components $a_1, \ldots, a_n$.

Matrix: $\mathbf{A} = \begin{bmatrix} a_{11} & \cdots & a_{1p} \\ \vdots & & \vdots \\ a_{n1} & \cdots & a_{np} \end{bmatrix} = (a_{ij})$ is an $n \times p$ matrix When $n = p$ $\mathbf{A}$ is called a square matrix of order n.

Transpose: the transpose of an $n \times p$ matrix $\mathbf{A}$ is a $p \times n$ matrix denoted by $\mathbf{A}'$ with

$$\mathbf{A}' = \begin{bmatrix} a_{11} & \cdots & a_{n1} \\ \vdots & & \vdots \\ a_{1p} & \cdots & a_{np} \end{bmatrix}.$$

Determinant of a square matrix $\mathbf{A}$: $|\mathbf{A}|$.

Trace of a square matrix $\mathbf{A}$: $\operatorname{tr} \mathbf{A} = \sum_{i=1}^{n} a_{ii}$.

Rank of $\mathbf{A}$: $\operatorname{rank}(\mathbf{A})$.

Identity matrix of order n: $\mathbf{I}_n$ or sometimes $\mathbf{I}$ for simplicity.

Zero matrix as well as zero vector: $\mathbf{0}$.

Inverse of a nonsingular square matrix $\mathbf{A}$: $\mathbf{A}^{-1}$.

Generalized inverse: A^- is a generalized inverse, or g-inverse for short, if $A = AA^-A$.

Kronecker product: $A \otimes B = (a_{ij}B)$. That is, the (i, j) element of A is replaced by the matrix $a_{ij}B$, forming the (i, j) block of $A \otimes B$.

Vector of ones: $\mathbf{1}' = (1, 1, \ldots, 1)$. A subscript n indicates the dimension of the vector of ones.

Signum function: $\mathrm{Sgn}(x) = 1$ for $x > 0$, for $x = 0$, and -1 for $x < 0$.

Real-valued random variables: These will be denoted by lower-case Latin italic letters (for example, x, y, r) and understood under circumstances.

Random vectors: These will be denoted by bold-faced lower-case Latin letters and the components of the vector by lower-case italic letters (for example, $\mathbf{x} = (x_1, \ldots, x_n)'$).

Random matrices: These will be denoted by bold-faced capital Latin letters and the entries of the matrix by lower-case Latin italic letters (for example, $\mathbf{X} = (x_{ij})$).

Expectations: The expectations of x, $\mathbf{x}$ and $\mathbf{X}$ are denoted by $E(x)$, $E(\mathbf{x}) = (E(x_1), \ldots, E(x_n))'$ and $E(\mathbf{X}) = (E(x_{ij}))$, respectively.

Variance of a random variable x: $\mathrm{Var}(x)$.

Covariance between x and y: $\mathrm{Cov}(x, y)$.

Covariance matrix of a random vector $\mathbf{x} = (x_1, \ldots, x_n)'$ will be denoted by $\mathrm{Cov}(\mathbf{x})$, where

$$\mathrm{Cov}(\mathbf{x}) = \begin{bmatrix} \mathrm{Var}(x_1) & \mathrm{Cov}(x_1, x_2) & \cdots & \mathrm{Cov}(x_1, x_n) \\ \mathrm{Cov}(x_2, x_1) & \mathrm{Var}(x_2) & \cdots & \mathrm{Cov}(x_2, x_n) \\ \vdots & \vdots & & \vdots \\ \mathrm{Cov}(x_n, x_1) & \mathrm{Cov}(x_n, x_2) & \cdots & \mathrm{Var}(x_n) \end{bmatrix}. \tag{1.11}$$

Moment of a random variable x of order α: $E(x^\alpha)$.

Mixed moment of a random vector $\mathbf{x}$: $E(x_1^{\alpha_1}, \ldots, x_n^{\alpha_n})$.

Cumulative distribution function (c.d.f.) of a random variable x:

$$F(t) = P(x \leqslant t).$$

Multivariate cumulative distribution function (m.c.d.f.) of a random vector $\mathbf{x}$:

$$F(\mathbf{t}) = F(x_1 \leqslant t_1, \ldots, x_n \leqslant t_n).$$

Probability density function (p.d.f.): $f(x)$ for x and $f(\mathbf{x})$ for $\mathbf{x}$.

Conditional distribution function of $\mathbf{x}$ given $\mathbf{y}$: $F(\mathbf{x}|\mathbf{y})$.

Characteristic function (c.f.): the c.f.s of $x, \mathbf{x}$ and $\mathbf{X}$ are respectively defined as:

$$\phi_x(t) = E(e^{itx}), \quad \phi_{\mathbf{x}}(\mathbf{t}) = E(e^{i\mathbf{t}'\mathbf{x}}),$$

and

$$\phi_{\mathbf{X}}(\mathbf{T}) = E(\exp(i \operatorname{tr} \mathbf{T}'\mathbf{x})),$$

where the matrices $\mathbf{T}$ and $\mathbf{X}$ are of the same size and $i = \sqrt{-1}$.

The following distributions will often be referred to in subsequent chapters:

1. The normal distribution with mean μ and variance σ^2: $N(\mu, \sigma^2)$; the p-dimensional multinormal distribution with mean vector $\boldsymbol{\mu}$ and covariance matrix $\boldsymbol{\Sigma}$: $N_p(\boldsymbol{\mu}, \boldsymbol{\Sigma})$.
2. Beta distribution: $\mathrm{Be}(\alpha, \beta)$ with the p.d.f.

$$\frac{1}{B(\alpha, \beta)} x^{\alpha-1}(1-x)^{\beta-1}, \quad \alpha, \beta > 0, \quad -1 \leqslant x \leqslant 1, \qquad (1.12)$$

   where $B(\alpha, \beta) = \Gamma(\alpha)\Gamma(\beta)/\Gamma(\alpha+\beta)$ is the beta function.
3. Gamma distribution: $\mathrm{Ga}(\alpha, \beta)$ with the p.d.f.

$$\begin{aligned} &\beta^{\alpha} x^{\alpha-1} e^{-\beta x}/\Gamma(\alpha), \quad \alpha, \beta > 0, \quad x \geqslant 0 \\ &\text{or} \quad \mathrm{Ga}(\alpha) \quad \text{if } \beta = 1. \end{aligned} \qquad (1.13)$$

4. Chi-square distribution with n degrees of freedom: χ_n^2.
5. Chi-distribution with m degrees of freedom: χ_m.
6. F-distribution with m and n degrees of freedom: $F(m, n)$.
7. t-distribution with n degrees of freedom: t_n.
8. Exponential distribution with parameter λ: $E(\lambda)$.

1.2.2 The $\overset{d}{=}$ operator

If x and y (or $\mathbf{x}$ and $\mathbf{y}$, or $\mathbf{X}$ and $\mathbf{Y}$) have the same distribution, we shall denote this fact by $x \overset{d}{=} y$ (or $\mathbf{x} \overset{d}{=} \mathbf{y}$ or $\mathbf{X} \overset{d}{=} \mathbf{Y}$ as the case may be).

In those situations when one can treat a problem directly in terms of random variables (rather than in terms of their characteristics) the operator $\overset{d}{=}$ plays a pivotal role. In this book most of the results will indeed be verified directly in terms of random variables (very often via a stochastic representation).

The following assertions (cf. Anderson and Fang, 1982, 1987, and Zolotarev, 1985) will be used in what follows:

1. If x is symmetrically distributed around the origin, then $x \stackrel{d}{=} -x$.
2. Assume that $\mathbf{x} \stackrel{d}{=} \mathbf{y}$ and $f_j(\cdot), j = 1, \ldots, m$, are measurable functions. Then
$$(f_1(\mathbf{x}), \ldots, f_m(\mathbf{x})) \stackrel{d}{=} (f_1(\mathbf{y}), \ldots, f_m(\mathbf{y})).$$
This assertion can be verified utilizing c.f.s.
3. If x, y and z are random variables such that $x \stackrel{d}{=} y$, it is not always the case that $z + x \stackrel{d}{=} z + y$. The latter, however, is valid provided z is independent of x and y respectively.
4. Assume that x, y and z are random variables, z is independent of x and y respectively. It is not always the case that $x + z \stackrel{d}{=} y + z$ implies $x \stackrel{d}{=} y$. We can approach this problem along the following lines: if $x + z \stackrel{d}{=} y + z$, then we have
$$\phi_x(t)\phi_z(t) = \phi_{x+z}(t) = \phi_{y+z}(t) = \phi_y(t)\phi_z(t), \tag{1.14}$$
If $\phi_z(t) \neq 0$ for almost all t, then $\phi_x(t) = \phi_y(t)$ for almost all t and $x \stackrel{d}{=} y$ by continuity of the c.f.s. If, however, $\phi_z(t) \neq 0$ only along $|t| < \delta$ for some $\delta > 0$, then
$$\phi_x(t) = \phi_y(t), \quad |t| < \delta. \tag{1.15}$$

Equality (1.15) cannot ensure that these two c.f.s are identical. Hsu (1954) gave a number of examples of this kind. He also introduced the following concept.

A c.f. $f(\cdot)$ is said to belong to the class E if for any c.f. $g(\cdot)$ such that $g(t) = f(t)$ in a neighbourhood of zero it is true that $f(t) = g(t)$ for all $t \in R^1$. A c.f. is said to belong to $\bar{E}$ if it does not belong to E.

Marcinkiewicz (1938) showed that the c.f. of a nonnegative (alternatively nonpositive) random variable belongs to E. Later Esseen (1945) proved that if a distribution has all finite moments and is determined uniquely by these moments, then the corresponding c.f. belongs to E.

If $\phi_x \in E$ (or $\phi_y \in E$), then the fact that (1.15) holds implies that (1.14) holds and $x \stackrel{d}{=} y$. Summarizing the above discussion we obtain the following lemma.

Lemma 1.1
Assume that x, y and z are random variables, and z is independent of x and y. The following statements are true:

(i) $x \stackrel{d}{=} y$ implies $x + z \stackrel{d}{=} y + z$;
(ii) if $\phi_z(t) \neq 0$, almost everywhere (a.e.) or $\phi_x \in E$, then $x + z \stackrel{d}{=} y + z$ implies that $x \stackrel{d}{=} y$.

Parallel results concerning products of independent variables are also available.

Lemma 1.2
Let x, y and z be random variables and z be independent of x and y. The following statements are true:

(i) $x \stackrel{d}{=} y$ implies $zx \stackrel{d}{=} zy$;
(ii) if $E(z^{it})$ and $E(|z|^{it} \operatorname{sgn}(z))$ are both nonzero for almost all real t, then $zx \stackrel{d}{=} zy$ implies that $x \stackrel{d}{=} y$.
(iii) if $P(z > 0) = 1$, the c.f. of $\{\log x | x > 0\}$ and the c.f. of $\{\log(-x) | x < 0\}$ belong to E, then $zx \stackrel{d}{=} zy$ implies that $x \stackrel{d}{=} y$.

The proof of (iii) and a special case of (ii) can be found in Anderson and Fang (1982, 1987). The proof of (ii) is given by Zolotarev (1985, p. 119 and p. 183), who developed **characteristic transform**, a powerful tool in studying products of independent random variables.

1.3 Groups and invariance

Groups and invariance have been applied in various branches of statistical methodology, especially in multivariate analysis. Recently, M. L. Eaton (1987) gave a series of lectures dealing with the concepts of invariance with special emphasis on applications in statistics. For our purposes, we need only a few very basic facts related to groups and invariance.

Definition 1.1
Let G be a set of transformations from a space $\mathscr{X}$ into itself which satisfy the following conditions:

(i) if $g_1 \in G$ and $g_2 \in G$, then $g_1 g_2 \in G$ where $g_1 g_2$ is defined as $(g_1 g_2)x = g_1(g_2 x)$ and $x \in \mathscr{X}$.
(ii) if $g \in G$, there exists a $g^{-1} \in G$ such that $gg^{-1} = g^{-1}g = e$, where e is the identity transformation in G.

A set of transformations G satisfying these two conditions will be called a **group**.

Definition 1.2
Two points x_1 and x_2 in $\mathscr{X}$ are said to be **equivalent** under G, if there exists a $g \in G$ such that $x_2 = gx_1$. We write $x_1 \sim x_2 \pmod G$.

Evidently, this equivalence relation has the following properties:

(i) $x \sim y \pmod G$;
(ii) $x \sim y \pmod G$ implies $y \sim x \pmod G$;
(iii) $x \sim y \pmod G$ and $y \sim z \pmod G$ implies $x \sim z \pmod G$.

Definition 1.3
A function $f(x)$ on $\mathscr{X}$ is said to be **invariant** under G if

$$f(gx) = f(x) \quad \text{for each } x \in \mathscr{X} \quad \text{and each } g \in G.$$

A function $f(x)$ on $\mathscr{X}$ is said to be a **maximal invariant** under G if it is invariant under G and if

$$f(x_1) = f(x_2) \quad \text{implies } x_1 \sim x_2 \pmod G.$$

Theorem 1.1
Assume $f(x)$ on $\mathscr{X}$ is a maximal invariant under G. Then a function $h(x)$ on $\mathscr{X}$ is invariant under G if and only if h is a function of $f(x)$.

PROOF If h is a function of $f(x)$, then $h(x) = v(f(x))$ for some $v \in G$ and all $x \in \mathscr{X}$. Evidently,

$$h(gx) = v(f(gx)) = v(f(x)) = h(x)$$

for all $g \in G$ and $x \in \mathscr{X}$, i.e. h is invariant.

Conversely, assume that h is invariant. Since f is a maximal invariant, $f(x_1) = f(x_2)$ implies $x_1 \sim x_2 \pmod G$, i.e. $x_2 = gx_1$ for some $g \in G$. Hence $h(x_2) = h(gx_1) = h(x_1)$ which shows that $h(x)$ depends on x only through $f(x)$. □

We shall now present two examples which will illustrate application of the definitions and results presented above.

Recall that $\mathcal{O}(n)$ denotes the group of $n \times n$ orthogonal matrices and $\mathcal{P}$ the group of permutation matrices, i.e. matrices $\mathbf{A}$ with the entries a_{ij} equal to 0 or 1 and $\mathbf{A} \in \mathcal{O}(n)$. In these groups the operation is matrix multiplication and they are often called the **orthogonal group** and the **permutation group** respectively.

Example 1.1
Let $\mathscr{X} = R^n$ and $G = \mathcal{P}$, the group of $n \times n$ permutation matrices. The action of $\mathbf{P} \in \mathcal{P}$ on $\mathbf{x} \in \mathscr{X}$ is $\mathbf{x} \to \mathbf{Px}$. A random vector $\mathbf{x} \in R^n$ is called exchangeable (cf. Section 1.1) if its distribution is invariant under G.

Example 1.2
Let $\mathscr{X} = R^n$ and $G = \mathcal{O}(n)$, the group of $n \times n$ orthogonal matrices. The action of $\mathbf{H} \in \mathcal{O}(n)$ on $\mathbf{x} \in R^n$ is $\mathbf{x} \to \mathbf{Hx}$. We want to show that $f(\mathbf{x}) = \mathbf{x}'\mathbf{x}$ is a maximal invariant under G. Indeed, f is invariant because $f(\mathbf{Hx}) = \mathbf{x}'\mathbf{H}'\mathbf{Hx} = \mathbf{x}'\mathbf{x}$ for all $\mathbf{x} \in R^n$ and $\mathbf{H} \in \mathcal{O}(n)$. Furthermore, if $\mathbf{x}_1'\mathbf{x}_1 = f(\mathbf{x}_1) = f(\mathbf{x}_2) = \mathbf{x}_2'\mathbf{x}_2$, there exists a $\mathbf{H} \in \mathcal{O}(n)$ such that $\mathbf{x}_2 = \mathbf{Hx}_1$, i.e. $\mathbf{x}_2 \sim \mathbf{x}_1$; thus $\mathbf{x}'\mathbf{x} = \|\mathbf{x}\|^2$ is a maximal invariant under G.

In general, given a group G, there may exist many maximal invariants under G. For example, $g(\mathbf{x}) = \|\mathbf{x}\|$, $h(\mathbf{x}) = a\|\mathbf{x}\|$, $a > 0$ are all maximal invariants under G in Example 1.2.

1.4 Dirichlet distribution

The Dirichlet distribution is a basic multivariate distribution for the purposes of this volume and it will often be referred to in the succeeding chapters. It is also applied nowadays with ever increasing frequency in statistical modelling, distribution theory and Bayesian inference (see e.g. Dickey and Chen, 1985; Aitchison, 1986). In this section a definition and discussion of this distribution is presented. We give detailed proofs in certain cases to acquaint the readers with some of the techniques to be used throughout this book.

Recall that a random variable y has a gamma distribution with parameter $\beta > 0$ (written as $y \sim \mathrm{Ga}(\beta)$) if y has the density function

$$(\Gamma(\beta))^{-1} y^{\beta-1} e^{-y}, \quad y > 0. \tag{1.16}$$

For $\beta = n/2$ in (1.16), $x = 2y$ has a chi-squared distribution with n degrees of freedom, i.e. $x \sim \chi_n^2$.

Definition 1.4

Let $x_1, \ldots, x_m, x_{m+1}$ be independent random variables. If $x_i \sim \mathrm{Ga}(\alpha_i)$ with $\alpha_i > 0$, $i = 1, \ldots, m, m+1$, set

$$y_j = x_j \Big/ \sum_{i=1}^{m+1} x_i, \quad j = 1, \ldots, m. \tag{1.17}$$

Then the distribution of $(y_1, \ldots, y_m)$ is called the **Dirichlet distribution** with parameters $\alpha_1, \ldots, \alpha_m, \alpha_{m+1}$. We shall write $(y_1, \ldots, y_m) \sim D_m(\alpha_1, \ldots, \alpha_m; \alpha_{m+1})$ or $(y_1, \ldots, y_{m+1}) \sim D_{m+1}(\alpha_1, \ldots, \alpha_{m+1})$.

From Definition 1.4, the density function of $(y_1, \ldots, y_m)$ can easily be obtained.

Theorem 1.2

The density function of $(y_1, \ldots, y_m) \sim D_m(\alpha_1, \ldots, \alpha_m; \alpha_{m+1})$ is given by

$$p_m(y_1, \ldots, y_m) = (B_m(\boldsymbol{\alpha}))^{-1} \prod_{i=1}^{m+1} y_i^{\alpha_i - 1},$$

$$\sum_{i=1}^{m+1} y_i = 1, \quad y_i \geqslant 0, \quad i = 1, \ldots, m+1, \tag{1.18}$$

where $B_m(\boldsymbol{\alpha}) = (\prod_{i=1}^{m+1} \Gamma(\alpha_i))/\Gamma(\alpha)$, $\alpha = \sum_{i=1}^{m+1} \alpha_i$ and the vector $\boldsymbol{\alpha} = (\alpha_1, \ldots, \alpha_{m+1})'$.

PROOF From (1.16) the joint density function of $(x_1, \ldots, x_{m+1})$ is

$$\frac{1}{\prod_{i=1}^{m+1} \Gamma(\alpha_i)} \prod_{i=1}^{m+1} x_i^{\alpha_i - 1} \exp\left(- \sum_{i=1}^{m+1} x_i \right).$$

Set

$$\begin{aligned} y_j &= x_j \Big/ \sum_{i=1}^{m+1} x_i, \quad j = 1, \ldots, m \\ y &= \sum_{i=1}^{m+1} x_i \end{aligned} \tag{1.19}$$

Then the Jacobian of transformation (1.19) is y^m. Therefore, the joint density function of $(y_1, \ldots, y_m, y)$ is

$$\frac{1}{\prod_{i=1}^{m+1} \Gamma(\alpha_i)} e^{-y} y^m y^{(\sum_{i=1}^{m+1} \alpha_i - m - 1)} \prod_{i=1}^{m} y_i^{\alpha_i - 1} \left(1 - \sum_{i=1}^{m} y_i\right)^{\alpha_{m+1} - 1}$$

$$= (B_m(\boldsymbol{\alpha}))^{-1} \prod_{i=1}^{m+1} y_i^{\alpha_i - 1} \{ y^{\alpha - 1} e^{-y} / \Gamma(\alpha) \}.$$

Since $y^{\alpha-1} e^{-y}/\Gamma(\alpha)$ is the density function of $y \sim \mathrm{Ga}(\alpha)$, the proof is completed by integrating out the density of y. □

It is easy to verify that for $m = 1$, $D_2(\alpha_1; \alpha_2)$ is the usual beta distribution $\mathrm{Be}(\alpha_1, \alpha_2)$.

The following theorems provide some properties of the Dirichlet distribution.

Theorem 1.3

Assume $\mathbf{y} = (y_1, \ldots, y_m)' \sim D_m(\alpha_1, \ldots, \alpha_m; \alpha_{m+1})$. Then for every $r_1, \ldots, r_m \geqslant 0$, we have

$$\mu_{r_1, \ldots r_m} = E\left(\prod_{j=1}^{m} y_j^{r_j}\right) = \left(\prod_{j=1}^{m} \alpha_j^{[r_j]}\right) \Big/ \alpha^{[r_1 + \cdots + r_m]} \tag{1.20}$$

where $x^{[r]} = x(x+1)\cdots(x+r-1)$ are descending factorials.

PROOF From (1.18) and $\Gamma(x+1) = x\Gamma(x)$, we have

$$\begin{aligned}
\mu_{r_1, \ldots, r_m} &= E\left(\prod_{j=1}^{m} y_j^{r_j}\right) \\
&= (B_m(\boldsymbol{\alpha}))^{-1} \int \left(1 - \sum_{j=1}^{m} y_j\right)^{\alpha_{m+1} - 1} \left(\prod_{j=1}^{m} y_j^{\alpha_j + r_j - 1} \, \mathrm{d}y_j\right) \\
&= (B_m(\boldsymbol{\alpha}))^{-1} \Gamma(\alpha_{m+1}) \prod_{j=1}^{m} \Gamma(\alpha_j + r_j) / \Gamma\left(\alpha + \sum_{j=1}^{m} r_j\right) \\
&= \left(\prod_{j=1}^{m} \alpha_j^{[r_j]}\right) \Big/ \alpha^{[r_1 + \cdots + r_m]}. \quad \square
\end{aligned}$$

Let $\mathbf{r} = (r_1, \ldots, r_m, 0)'$. Then (1.20) can be expressed as

$$\mu_{r_1, \ldots, r_m} = B_m(\boldsymbol{\alpha} + \mathbf{r}) / B_m(\boldsymbol{\alpha}). \tag{1.21}$$

In particular,

$$E(y_j) = \alpha_j/\alpha \quad \text{and} \quad \mathrm{Var}(y_j) = \frac{\alpha_j(\alpha - \alpha_j)}{\alpha^2(\alpha + 1)} \tag{1.22}$$

and the covariance between y_i and y_j is

$$-\alpha_i\alpha_j/\alpha^2(\alpha + 1). \tag{1.23}$$

Note that the covariance is a negative number, which indicates the negative association between the components $y_i (i = 1, \ldots, m)$.

The Dirichlet distribution has an interesting amalgamation property: if the $n = m + 1$ components $x_1, \ldots, x_n$, $\sum_{i=1}^n x_i = 1$, of a $D_n(\alpha_1, \ldots, \alpha_n)$ distribution in the symmetric form are lumped into k components w_i, $\sum_{i=1}^k w_i = 1$, this amalgamated vector $\mathbf{w} = (w_1, \ldots, w_k)$ has a Dirichlet distribution $D_k(\alpha_1^*, \ldots, \alpha_k^*)$ where α_i^* is the sum of the α_j's corresponding to those components which add up to w_i. This is part of the following theorem which is in terms of the symmetric form of a Dirichlet distribution.

Theorem 1.4

Let $\mathbf{x} \sim D_n(\boldsymbol{\alpha})$ be partitioned into k subvectors $\mathbf{x}^{(1)}, \ldots, \mathbf{x}^{(k)}$ and $\boldsymbol{\alpha}$ into the corresponding subvectors $\boldsymbol{\alpha}^{(1)}, \ldots, \boldsymbol{\alpha}^{(k)}$. Let w_i and α_i^* be respectively the sums of the components of $\mathbf{x}^{(i)}$ and $\boldsymbol{\alpha}^{(i)}$, and $\mathbf{y}^{(i)} = \mathbf{x}^{(i)}/w_i$. The following statements are true:

(i) The vectors $\mathbf{y}^{(i)}, \ldots, \mathbf{y}^{(k)}$, and $\mathbf{w}$ are mutually independent.
(ii) $\mathbf{w}$ is distributed as $D_k(\alpha_1^*, \ldots, \alpha_k^*)$.
(iii) $\mathbf{y}^{(i)}$ is distributed as $D_{n_i}(\boldsymbol{\alpha}^{(i)})$, $i = 1, \ldots, k$.

PROOF Writing $\mathbf{x}^{(i)} = (\mathbf{x}_{i1}, \ldots, x_{in_i})'$ and substituting $x_{ij} = w_i y_{ij}$, where $i = 1, \ldots, k$ and $j = 1, \ldots, n_i$, into the density function of $\mathbf{x}$, we have

$$\begin{aligned} f(\mathbf{x}) &= \frac{\Gamma(\sum\sum \alpha_{ij})}{\prod_{i=1}^{k} \sum_{j=1}^{n_i} \Gamma(\alpha_{ij})} \prod_{i=1}^{k} \prod_{j=1}^{n_i} x_{ij}^{\alpha_{ij}-1} \\ &= \prod_{i=1}^{k} \left(\frac{\Gamma(\alpha_i^*)}{\prod_j \Gamma(\alpha_{ij})} \prod_{j=1}^{n_i} y_{ij}^{\alpha_{ij}-1} \right) \cdot \frac{\Gamma(\alpha)}{\prod_i \Gamma(\alpha_i^*)} \prod_{i=1}^{k} w_i^{\alpha_i^* - n_i}, \end{aligned}$$

where $\alpha = \sum\sum \alpha_{ij} = \sum \alpha_i^*$. The following Jacobians are obvious:

$$J(\mathbf{x}^{(1)} \to (w_1, y_{11}, \ldots, y_{1n_1-1})) = w_1^{n_1-1},$$
$$J(\mathbf{x}^{(2)} \to (w_2, y_{21}, \ldots, y_{2n_2-1})) = w_2^{n_2-1},$$
$$\vdots$$
$$J(x_{k1}, \ldots, x_{kn_k-1}) \to (y_{k1}, \ldots, y_{kn_k-1})) = w_k^{n_k-1}.$$

Thus the resultant Jacobian is $\prod_{i=1}^{k} w_i^{n_i-1}$. Substituting this Jacobian into the joint density of $\mathbf{w}$ and $\mathbf{y}^{(1)}, \ldots, \mathbf{y}^{(k)}$, we arrive at the conclusions. □

From the above theorem it follows immediately that all the marginal distributions are Dirichlet (or beta) distributions. We state this result as the following theorem.

Theorem 1.5
If $(y_1, \ldots, y_m) \sim D_m(\alpha_1, \ldots, \alpha_m; \alpha_{m+1})$ then for any $s < m$, we have

$$(y_1, \ldots, y_s) \sim D_s(\alpha_1, \ldots, \alpha_s; \alpha_{s+1}^*), \tag{1.24}$$

where $\alpha_{s+1}^* = \alpha_{s+1} + \cdots + \alpha_m + \alpha_{m+1}$.
In particular, for any i ($i = 1, \ldots, m$), we have

$$y_i \sim \mathrm{Be}(\alpha_i, \alpha - \alpha_i).$$

From the density of the Dirichlet distribution we easily obtain all the marginal densities of the random variable $\mathbf{x}/\|\mathbf{x}\|$, where $\mathbf{x} \sim N_n(\mathbf{0}, \mathbf{I}_n)$. Since

$$(x_1^2/\|\mathbf{x}\|^2, \ldots, x_k^2/\|\mathbf{x}\|^2) \sim D_k(1/2, \ldots, 1/2; (n-k)/2) \quad \text{for } 0 < k < n;$$

this implies that the density of $(|x_1|/\|\mathbf{x}\|, \ldots, |x_k|/\|\mathbf{x}\|)$ is

$$\frac{\Gamma(n/2)2^k}{\Gamma((n-k)/2)\pi^{k/2}}\left(1 - \sum_{i=1}^{k} x_i^2\right)^{(n-k)/2-1}, \tag{1.25}$$

$$x_i \geqslant 0, \quad i = 1, \ldots, k, \quad \sum_{i=1}^{k} x_i^2 < 1$$

and the density of $(x_1/\|\mathbf{x}\|, \ldots, x_k/\|\mathbf{x}\|)$ is

$$\frac{\Gamma(n/2)}{\Gamma((n-k)/2)\pi^{k/2}}\left(1 - \sum_{i=1}^{k} x_i^2\right)^{(n-k)/2-1}, \qquad \sum_{i=1}^{k} x_i^2 < 1. \tag{1.26}$$

Formula (1.26) will be used in Section 2.2.
The following result is easily verified by standard calculations (see Problem 1.3 at the end of this chapter).

Theorem 1.6
Assume that $(y_1,\ldots,y_m) \sim D_m(\alpha_1,\ldots,\alpha_m;\alpha_{m+1})$. Set

$$y_j^* = y_j \Big/ \left(1 - \sum_{i=1}^{s} y_i\right), \quad j = s+1,\ldots,m,$$

then

$$(y_{s+1}^*,\ldots,y_m^* \mid y_1,\ldots,y_s) \sim D_{m-s}(\alpha_{s+1},\ldots,\alpha_m;\alpha_{m+1}).$$

Theorems 1.5 and 1.6 are also presented in Johnson and Kotz (1972). From Theorem 1.2 we have

$$B_m(\boldsymbol{\alpha}) = \int_{D(x_1,\ldots,x_m)} \prod_{i=1}^{m} x_i^{\alpha_i - 1} \left(1 - \sum_{i=1}^{m} x_i\right)^{\alpha_{m+1}-1} \mathrm{d}x_1 \cdots \mathrm{d}x_m, \tag{1.27}$$

where

$$D(x_1,\ldots,x_m) = \left\{(x_1,\ldots,x_m)\colon x_i \geqslant 0, i = 1,\ldots,m, \sum_{i=1}^{m} x_i < 1\right\}. \tag{1.28}$$

The integral (1.27) is attributed to P.G.L. Dirichlet (1805–59). Joseph Liouville (1809–82) generalized it as stated in Lemma 1.3 below. It is pivotal in the developments of Liouville distributions which we shall discuss in Chapter 6. For any nonnegative measurable function $f(\cdot)$, let

$$I_m(f \mid \boldsymbol{\alpha}) = \int_{R_+^m} f\left(\sum_{i=1}^{m} x_i\right)\left(\prod_{i=1}^{m} x_i^{\alpha_i - 1}\,\mathrm{d}x_i\right), \tag{1.29}$$

where $\boldsymbol{\alpha} = (\alpha_1,\ldots,\alpha_m)'$ and $R_+^m = \{(x_1,\ldots,x_m)\colon x_i \geqslant 0, i = 1,\ldots,m\}$.

Lemma 1.3
It is true that

$$I_m(f \mid \boldsymbol{\alpha}) = B_{m-1}(\boldsymbol{\alpha}) \cdot I_1\left(f \,\Big|\, \sum_{i=1}^{m} \alpha_i\right), \tag{1.30}$$

provided that either integral exists.

PROOF Transforming

$$\begin{cases} y_i = x_i, & i = 1,\ldots,m-1 \\ y_m = x_1 + \cdots + x_m \end{cases}$$

in (1.29) and noting that the Jacobian of the transformation is 1, we have

$$I_m(f\,|\,\boldsymbol{\alpha}) = \int_D f(y_m)\left(y_m - \sum_{i=1}^{m-1} y_i\right)^{\alpha_m - 1}\left(\prod_{i=1}^{m-1} y_i^{\alpha_i - 1}\,\mathrm{d}y_i\right),$$

where

$$D = \left\{(y_1,\ldots,y_m)\colon y_i \geqslant 0,\, i = 1,\ldots,m, \sum_{i=1}^{m-1} y_i \leqslant y_m\right\}.$$

Consider now yet another transformation $u_i = y_i/y_m$, $i = 1,\ldots,m-1$, $y = y_m$. Its Jacobian is y^{m-1}, thus

$$I_m(f\,|\,\boldsymbol{\alpha}) = \int_{D(u_1,\ldots,u_{m-1})}\left(1 - \sum_{i=1}^{m-1} u_i\right)^{\alpha_m - 1}\left(\prod_{i=1}^{m-1} u_i^{\alpha_i - 1}\,\mathrm{d}u_i\right)$$
$$\times \int_0^\infty f(y)\, y^{\sum_{i=1}^m \alpha_i - 1}\,\mathrm{d}y$$

which completes the proof. □

This integral can be extended to the entire R^m as follows (see Zhang and Fang, 1982).

Lemma 1.4
For $\alpha_i > 0$, $i = 1,\ldots,m$, we have

$$\int_{R^m} f\left(\sum_{i=1}^m x_i^2\right)\left(\prod_{i=1}^m |x_i|^{2\alpha_i - 1}\,\mathrm{d}x_i\right) = I_m(f\,|\,\boldsymbol{\alpha}). \tag{1.31}$$

PROOF By the symmetry of $x_1,\ldots,x_m$ about the origin we have

$$I = \int f\left(\sum_{i=1}^m x_i^2\right)\left(\prod_{i=1}^m |x_i|^{2\alpha_i - 1}\,\mathrm{d}x_i\right)$$
$$= 2^m \int_{R_+^m} f\left(\sum_{i=1}^m x_i^2\right)\prod_{i=1}^m x_i^{2\alpha_i - 1}\,\mathrm{d}x_i.$$

Set $u_i = x_i^2$, $i = 1,\ldots,m$, then the Jacobian of this transformation is $(2^m\prod_{i=1}^m u_i^{1/2})^{-1}$. Hence we have

$$I = \int_{R_+^m} f\left(\sum_{i=1}^m u_i\right)\left(\prod_{i=1}^m u_i^{\alpha_i - 1}\,\mathrm{d}u_i\right) = I_m(f\,|\,\boldsymbol{\alpha}). \quad \square$$

Setting $\alpha_i = 1/2$, $i = 1, \ldots, m$ in (1.31) we immediately obtain the following corollary.

Corollary

$$\int f\left(\sum_{i=1}^{m} x_i^2\right) \mathrm{d}x_1 \cdots \mathrm{d}x_m = \frac{\pi^{m/2}}{\Gamma(m/2)} \int_0^\infty y^{m/2-1} f(y) \mathrm{d}y$$
$$= \pi^{m/2} (\Gamma(m/2))^{-1} I_1(f \mid m/2). \quad (1.32)$$

where $I_1(f \mid m/2)$ is given by (1.29).

Formula (1.32) has numerous applications.

Example 1.3
Substituting

$$f(y) = \begin{cases} (2\pi)^{-m/2} e^{-y/2} & \text{if } y \leqslant x \\ 0 & \text{if } y > x \end{cases}$$

into (1.32), we have

$$\int_{\sum_{i=1}^m x_i^2 \leqslant x} (2\pi)^{-m/2} e^{-(x_1^2 + \cdots + x_m^2)/2} \mathrm{d}x_1 \cdots \mathrm{d}x_m$$
$$= \pi^{m/2} (\Gamma(m/2))^{-1} \int_0^x (2\pi)^{-m/2} e^{-y/2} y^{m/2-1} \mathrm{d}y$$
$$= 2^{-m/2} (\Gamma(m/2))^{-1} \int_0^x y^{m/2-1} e^{-y/2} \mathrm{d}y.$$

This implies that if $\mathbf{x} \sim N_m(\mathbf{0}, \mathbf{I}_m)$, then the distribution function of $Y = \mathbf{x}'\mathbf{x}$ is given by the formula above. We have thus established the well-known fact that the density of the chi-squared distribution with m degrees of freedom is

$$(2^{m/2} \Gamma(m/2))^{-1} y^{m/2-1} e^{-y/2}. \quad (1.33)$$

Example 1.4
Substituting

$$f(y) = \begin{cases} 1 & \text{if } y \leqslant a^2 \\ 0 & \text{if } y > a^2 \end{cases}$$

into (1.32), we obtain a formula for the volume of an m-dimensional sphere of radius a:

$$V_m(a) = \int_{\sum_{i=1}^m x_i^2 \leqslant a^2} \mathrm{d}x_1 \cdots \mathrm{d}x_m = \pi^{m/2}(\Gamma(m/2))^{-1} \int_0^{a^2} y^{m/2-1}\mathrm{d}y$$

$$= \pi^{m/2}(\Gamma((m+2)/2))^{-1} a^m = \frac{2\pi^{m/2}}{m\Gamma(m/2)} a^m.$$

The reader can find more properties of Dirichlet distributions in Aitchison (1986).

Problems

1.1 Show that if $x_i \sim \chi^2_{n_i}$, $i = 1, \ldots, m, m+1$, are independent and $x = \sum_{i=1}^{m+1} x_i$ then

(i) $\left(\dfrac{x_1}{x}, \ldots, \dfrac{x_m}{x}\right) \sim D(n_1/2, \ldots, n_m/2; n_{m+1}/2)$;

(ii) $x \sim \chi^2_n$ is independent of $\left(\dfrac{x_1}{x}, \ldots, \dfrac{x_m}{x}\right)$ with $n = \sum_{i=1}^{m+1} n_i$.

1.2 Partition $\mathbf{x} \sim N_n(\mathbf{0}, \mathbf{I}_n)$ into $m+1$ parts $\mathbf{x}^{(1)}, \ldots, \mathbf{x}^{(m)}, \mathbf{x}^{(m+1)}$ with $n_1, \ldots, n_m, n_{m+1}$ components respectively, and let $\|\mathbf{x}\|^2 = \mathbf{x}'\mathbf{x}$. Verify that

(i) $\left(\dfrac{\|\mathbf{x}^{(1)}\|^2}{\|\mathbf{x}\|^2}, \ldots, \dfrac{\|\mathbf{x}^{(m)}\|^2}{\|\mathbf{x}\|^2}\right) \sim D_m(n_1/2, \ldots, n_m/2; n_{m+1}/2)$;

(ii) $\|\mathbf{x}\|^2$ is independent of $\left(\dfrac{\|\mathbf{x}^{(1)}\|^2}{\|\mathbf{x}\|^2}, \ldots, \dfrac{\|\mathbf{x}^{(m)}\|^2}{\|\mathbf{x}\|^2}\right)$.

1.3 Prove Theorem 1.6, which is related to the conditional distribution of Dirichlet random variables.
Hint: Derive the joint distribution of $(y_1, \ldots, y_s | y^*_{s+1}, \ldots, y^*_m)$.

1.4 Let f be a nonnegative integrable function. Verify the following formulas:

(i) $$\int_{R^m_+} f\left(\sum_1^m x_i\right) \prod_1^m (x_i^{\alpha_i - 1} \mathrm{d}x_1) = \frac{\prod_1^m \Gamma(\alpha_i)}{\Gamma(\alpha)} \int_0^\infty f(y) y^{\alpha-1}\,\mathrm{d}y,$$

where $\alpha = \sum_1^m \alpha_i$.

(ii) $$\int_{R_+^m} f\left(\sum_1^m x_i\right) \mathrm{d}x_1 \cdots \mathrm{d}x_m = \frac{1}{(m-1)!} \int_0^\infty y^{m-1} f(y) \mathrm{d}y.$$

1.5 Assume that $\mathbf{y} = (y_1, \ldots, y_m)' \sim D_m(\alpha_1, \ldots, \alpha_m; \alpha_{m+1})$ and $\psi(t) = \mathrm{d} \log \Gamma(t)/\mathrm{d}t$ and $\psi'(t)$ $(t > 0)$ are the digamma and trigamma functions, respectively. Then

(i) $E\{\log(y_i/y_j)\} = \psi(\alpha_i) - \psi(\alpha_j)$;

(ii) $\mathrm{Var}\{\log(y_i/y_j)\} = \psi'(\alpha_i) + \psi'(\alpha_j)$;

(iii) $\mathrm{Cov}\{\log(y_i/y_k, \log(y_j/y_k)\} = \psi'(\alpha_k)$.

1.6 Let S_n be the area of unit sphere surface in R^n. Verify that

$$S_n = \frac{2\pi^{n/2}}{\Gamma(n/2)}.$$

CHAPTER 2

Spherically and elliptically symmetric distributions

In this chapter classes of spherically and elliptically symmetric distributions which are extensions of multinormal distributions are defined and a discussion of those classes is given. Some basic properties are presented in Sections 2.2, 2.3, 2.4 and 2.5. In Section 2.6, we concentrate on the class of mixtures of normal distributions, the most important subclass of elliptically symmetric distributions. Some applications to robust statistics and linear models are given in Section 2.7. Finally log-elliptically symmetric and additive logistic elliptical distributions are briefly discussed in Section 2.8 and complex elliptically symmetric distributions in Section 2.9.

2.1 Introduction and definition

The subject matter of the present chapter is a comprehensive study of the whole range of spherically and elliptically symmetric distributions which is to the best of our knowledge the first one in monographic literature. Although – as indicated in the excellent review and bibliography of elliptically symmetric distributions by Chmielewski (1981) – earlier papers on this topic can be traced to Maxwell (1860), Bartlett (1934) and Hartman and Wintner (1940), the modern era of research in this field starts perhaps with the engineering applications considered by Blake and Thomas (1968) and McGraw and Wagner (1968), followed by Chu (1973), and more prominently for statisticians with Dempster (1969) and the first organized presentation of this family by Kelker (1970). Since then we are witnessing a substantial well-organized and ever increasing amount of research of properties and applications of these distributions. A landmark paper by Cambanis, Huang and Simons (1981) presents a systematic treatment for elliptical distributions. Another review paper (in Chinese) on this topic by Fang (1987) is influential and prominent among specialists and practitioners in China. An admirably lucid discussion of 'elliptically contoured distributions' is presented in the recent book by M.E. Johnson (1987, Ch. 6). The report by Diaconis and Freedman (1986) and the abstract by Diaconis (1987) claiming (and proving) that elliptically contoured distributions are a 'minor modification' of independent identically distributed normal may be also of interest for a different perspective. Our presentation of these distributions will in a sense be an extensive supplement to the one given in Johnson's book, using a more general approach. We shall be emphasizing the mathematical rigour and concentrating on the results not mentioned or only briefly sketched in his book.

There are several ways to define spherically and elliptically symmetric distributions. A spherical distribution is an extension of the standard multinormal distribution $N(\mathbf{0}, \mathbf{I}_n)$ and an elliptical distribution is an extension of $N(\boldsymbol{\mu}, \boldsymbol{\Sigma})$. It is well known that the multivariate normal distribution $N(\boldsymbol{\mu}, \boldsymbol{\Sigma})$ can be defined by the standard normal distribution $N(\mathbf{0}, \mathbf{I})$ as follows

$$\mathbf{x} \stackrel{d}{=} \boldsymbol{\mu} + \mathbf{A}'\mathbf{y}, \tag{2.1}$$

where $\mathbf{x} \sim N(\boldsymbol{\mu}, \boldsymbol{\Sigma})$, $\mathbf{y} \sim N(\mathbf{0}, \mathbf{I})$ and $\boldsymbol{\Sigma} = \mathbf{A}'\mathbf{A}$. It is easy to find many properties of $N(\boldsymbol{\mu}, \boldsymbol{\Sigma})$ parallel to that of $N(\mathbf{0}, \mathbf{I})$ through (2.1). We shall therefore define and investigate the spherical distribution first. Our first definition characterizes a spherical distribution in an intuitively appealing way.

Definition 2.1
An $n \times 1$ random vector $\mathbf{x}$ is said to have a **spherically symmetric** distribution (or simply **spherical** distribution) if for every $\boldsymbol{\Gamma} \in \mathcal{O}(n)$,

$$\boldsymbol{\Gamma}\mathbf{x} \stackrel{d}{=} \mathbf{x} \tag{2.2}$$

where the sign $\stackrel{d}{=}$ has been defined in Section 1.2, and $\mathcal{O}(n)$ denotes the set of $n \times n$ orthogonal matrices. The set $\mathcal{O}(n)$ is a group (cf. Section 1.3), called the orthogonal group, with the group operation being the ordinary matrix multiplication.

Theorem 2.1
An n-vector $\mathbf{x}$ has a spherical distribution if and only if its characteristic function (c.f.) $\psi(\mathbf{t})$ satisfies one of the following

equivalent conditions:

(i) $\psi(\mathbf{\Gamma}'\mathbf{t}) = \psi(\mathbf{t})$ for any $\mathbf{\Gamma} \in \mathcal{O}(n)$;
(ii) There exists a function $\phi(\cdot)$ of a scalar variable such that $\psi(\mathbf{t}) = \phi(\mathbf{t}'\mathbf{t})$.

PROOF Note that for any square matrix $\mathbf{A}$, the c.f. of $\mathbf{Ax}$ equals $\psi(\mathbf{A}'\mathbf{t})$, that is,

$$E(e^{i\mathbf{t}'\mathbf{Ax}}) = E(e^{i(\mathbf{A}'\mathbf{t})'\mathbf{x}}) = \psi(\mathbf{A}'\mathbf{t}).$$

Thus (i) is equivalent to (2.2). Now (ii) implies (i), since

$$\psi(\mathbf{\Gamma}'\mathbf{t}) = \phi((\mathbf{\Gamma}'\mathbf{t})'(\mathbf{\Gamma}'\mathbf{t})) = \phi(\mathbf{t}'\mathbf{\Gamma}\mathbf{\Gamma}'\mathbf{t}) = \phi(\mathbf{t}'\mathbf{t}) = \psi(\mathbf{t}).$$

Conversely, (i) implies that $\psi(\mathbf{t})$ is an invariant function w.r.t. the group $\mathcal{O}(n)$ which has the maximal invariant $\mathbf{t}'\mathbf{t}$. Thus $\psi(\mathbf{t})$ must be a function of $\mathbf{t}'\mathbf{t}$ (cf. Section 1.3), implying (ii). □

From now on, we shall write $\mathbf{x} \sim S_n(\phi)$ to mean that $\mathbf{x}$ has a c.f. of the form $\phi(\mathbf{t}'\mathbf{t})$, where $\phi(\cdot)$ is a function of a scalar variable, called the **characteristic generator** of the spherical distribution. This concept is widely used in the sequel.

Example 2.1
Let $\mathbf{u}^{(n)}$ denote a random vector distributed uniformly on the unit sphere surface in R^n. It is quite obvious that $\mathbf{\Gamma u}^{(n)} \stackrel{d}{=} \mathbf{u}^{(n)}$ for every $\mathbf{\Gamma} \in \mathcal{O}(n)$ and is distributed according to a spherical distribution.

Example 2.2
Let $\mathbf{x}$ denote a random vector distributed uniformly inside the unit sphere in R^n. In accordance with Definition 2.1, $\mathbf{x}$ has a spherical distribution.

Example 2.3 (Multinormal distribution)
Let $\mathbf{x}' = (x_1, \ldots, x_n)$ be distributed as $N_n(\mathbf{0}, \mathbf{I})$. Since the c.f. of x_1 is $\exp(-t_1^2/2)$, the c.f. of $\mathbf{x}$ is

$$\exp\{-\tfrac{1}{2}(t_1^2 + \cdots + t_n^2)\} = \exp\{-\tfrac{1}{2}\mathbf{t}'\mathbf{t}\}.$$

Hence from Theorem 2.1 $\mathbf{x}$ has a spherical distribution $S_n(\phi)$ with characteristic generator $\phi(u) = \exp(-u/2)$.

Theorem 2.1 shows that the characteristic function of a spherical distribution is generated by some scalar function. A natural question is: what are these characteristic generators? To answer this question, we need to introduce the following notation.

Denote the family of all possible characteristic generators for an $n \times 1$ random vector by

$$\Phi_n = \{\phi(\cdot): \phi(t_1^2 + \cdots + t_n^2) \text{ is an } n\text{-dimensional c.f.}\}. \tag{2.3}$$

It is easy to see that $\Phi_1 \supset \Phi_2 \supset \Phi_3 \supset \cdots$.

Let

$$\Phi_\infty = \bigcap_{n=1}^{\infty} \Phi_n. \tag{2.4}$$

Theorem 2.2

A function $\phi \in \Phi_n$ if and only if

$$\phi(x) = \int_0^\infty \Omega_n(xr^2)\mathrm{d}F(r) \tag{2.5}$$

where $F(\cdot)$ is a c.d.f. over $[0, \infty)$ and

$$\Omega_n(\mathbf{y}'\mathbf{y}) = \int_{S:\mathbf{x}'\mathbf{x}=1} e^{i\mathbf{y}'\mathbf{x}}\mathrm{d}S/S_n, \tag{2.6}$$

where S_n is the area of unit sphere surface in R^n (cf. Problem 1.6), i.e. $\Omega_n(\mathbf{t}'\mathbf{t})$ is the c.f. of $\mathbf{u}^{(n)}$ (cf. Example 2.1).

PROOF Assume $\phi(\cdot) \in \Phi_n$, then $g(t_1, \ldots, t_n) \equiv \phi(\mathbf{t}'\mathbf{t})$ is a c.f. of some random vector $\mathbf{y}$ with a c.d.f. $G(\mathbf{y})$. Thus $g(t_1, \ldots, t_n)$ is a symmetric function of $t_1, \ldots, t_n$. For every $\mathbf{x}$ with $\mathbf{x}'\mathbf{x} = \|\mathbf{x}\|^2 = 1$, we have

$$g(t_1, \ldots, t_n) = \phi(\mathbf{t}'\mathbf{t}) = g(\|\mathbf{t}\| x_1, \ldots, \|\mathbf{t}\| x_n)$$

and

$$\begin{aligned}
\phi(\mathbf{t}'\mathbf{t}) &= \frac{1}{S_n}\int_{S:\|\mathbf{x}\|=1} g(\|\mathbf{t}\| x_1, \ldots, \|\mathbf{t}\| x_n)\mathrm{d}S \\
&= \frac{1}{S_n}\int_{S:\|\mathbf{x}\|=1}\left[\int_{R^n} e^{i\|\mathbf{t}\|\mathbf{x}'\mathbf{y}}\mathrm{d}G(\mathbf{y})\right]\mathrm{d}S \\
&= \int_{R^n}\left[\frac{1}{S_n}\int_{S:\|\mathbf{x}\|=1} e^{i\|\mathbf{t}\|\mathbf{x}'\mathbf{y}}\mathrm{d}S\right]\mathrm{d}G(\mathbf{y}) \\
&= \int_{R^n}\Omega_n(\|\mathbf{t}\|^2\|\mathbf{y}\|^2)\mathrm{d}G(\mathbf{y}).
\end{aligned}$$

Let

$$F(u) = \int_{\|\mathbf{y}\| \leqslant u} \mathrm{d}G(\mathbf{y})$$

then $F(\cdot)$ is a c.d.f. over $[0, \infty)$ and

$$\phi(x) = \int_0^\infty \Omega_n(xu^2)\mathrm{d}F(u).$$

Assuming that $\phi(\cdot)$ can be expressed in the form (2.5), take a random variable $r \sim F(x)$ to be independent of $\mathbf{u}^{(n)}$, then the c.f. of $r\mathbf{u}^{(n)}$ is

$$E(e^{it'r\mathbf{u}^{(n)}}) = \int_0^\infty E(e^{irt'\mathbf{u}^{(n)}})\,\mathrm{d}F(r)$$

$$= \int_0^\infty \Omega_n(r^2 \|\mathbf{t}\|^2)\,\mathrm{d}F(r) = \phi(\|\mathbf{t}\|^2)$$

i.e. $\phi(\|\mathbf{t}\|^2)$ is the c.f. of $r\mathbf{u}^{(n)}$ and $\phi(\cdot) \in \Phi_n$. □

This theorem is due to Schoenberg (1938). The following important corollary is a consequence of Theorem 2.2. For an alternative representation, see Dickey and Chen (1987).

Corollary

Let the c.f. of an $n \times 1$ random vector $\mathbf{x}$ be $\phi(\mathbf{t}'\mathbf{t})$ and $\phi \in \Phi_n$. Then $\mathbf{x}$ has a stochastic representation

$$\mathbf{x} \stackrel{d}{=} r\mathbf{u}^{(n)} \tag{2.7}$$

where r is independent of $\mathbf{u}^{(n)}$, and $r \sim F(x)$ is related to ϕ by the relation (2.5).

From now on, whenever we write $\mathbf{x} \stackrel{d}{=} r\mathbf{u}^{(n)}$ for $\mathbf{x} \sim S_n(F)$, it will mean that r and $\mathbf{u}^{(n)}$ have the same meaning as in this corollary and $r \sim F(x)$. We shall call r the **generating variate**, F the **generating c.d.f.**, and $\mathbf{u}^{(n)}$ the **uniform base** of the spherical distribution. We shall write $\mathbf{x} \sim S_n^+(\phi)$ to mean that $\mathbf{x} \sim S_n(\phi)$ and $P(\mathbf{x} = \mathbf{0}) = 0$.

Theorem 2.3

Suppose $\mathbf{x} \stackrel{d}{=} r\mathbf{u}^{(n)} \sim S_n^+(\phi)$, then

$$\|\mathbf{x}\| \stackrel{d}{=} r, \qquad \mathbf{x}/\|\mathbf{x}\| \stackrel{d}{=} \mathbf{u}^{(n)}. \tag{2.8}$$

Moreover, $\|\mathbf{x}\|$ and $\mathbf{x}/\|\mathbf{x}\|$ are independent.

PROOF Since $P(\mathbf{x}=\mathbf{0})=0=P(r=0)$, choose $f_1(\mathbf{x})=(\mathbf{x}'\mathbf{x})^{1/2}$ and $f_2(\mathbf{x})=\mathbf{x}/(\mathbf{x}'\mathbf{x})^{1/2}$. Then from assertion 2 in Section 1.2.2

$$\begin{bmatrix} \|\mathbf{x}\| \\ \mathbf{x}/\|\mathbf{x}\| \end{bmatrix} = \begin{bmatrix} f_1(\mathbf{x}) \\ f_2(\mathbf{x}) \end{bmatrix} \stackrel{d}{=} \begin{bmatrix} f_1(r\mathbf{u}^{(n)}) \\ f_2(r\mathbf{u}^{(n)}) \end{bmatrix} = \begin{bmatrix} r \\ \mathbf{u}^{(n)} \end{bmatrix}.$$

This completes the proof. □

Theorem 2.4
Let $\mathbf{x}=(x_1,\ldots,x_n)'$ be an $n\times 1$ random vector. Then $\mathbf{x}$ is a spherical random vector if and only if for any $\mathbf{a}\in R^n$, we have

$$\mathbf{a}'\mathbf{x} = \|\mathbf{a}\| x_1. \tag{2.9}$$

PROOF If $\mathbf{x}\sim S_n(\phi)$, the c.f. of x_1 is $\phi(t^2)$. Then $E(\exp(it\mathbf{a}'\mathbf{x}))=\phi(t^2\mathbf{a}'\mathbf{a})=\phi(t^2\|\mathbf{a}\|^2)=E(\exp(it\|\mathbf{a}\|x_1))$, which implies (2.9). Conversely, if (2.9) holds for any $\mathbf{a}\in R^n$ we have

$$E(\exp(i\mathbf{a}'\mathbf{x}))=E(\exp(i\|\mathbf{a}\|x_1))=\phi(\|\mathbf{a}\|),$$

where $\phi(t)$ is the c.f. of x_1. From Theorem 2.1(ii) $\mathbf{x}$ is spherical. □

The above results are summarized in the following theorem.

Theorem 2.5
Let $\mathbf{x}=(x_1,\ldots,x_n)'$ be an $n\times 1$ random vector. The following statements are equivalent:

(i) $\mathbf{x}\stackrel{d}{=}\mathbf{\Gamma x}$ for every $\mathbf{\Gamma}\in\mathcal{O}(n)$;
(ii) The c.f. of $\mathbf{x}$ is of the form $\phi(\mathbf{t}'\mathbf{t})$, for some $\phi\in\mathbf{\Phi}_n$;
(iii) $\mathbf{x}$ has a stochastic representation $\mathbf{x}\stackrel{u}{=}r\mathbf{u}^{(n)}$ for some $r\geqslant 0$ which is independent of $\mathbf{u}^{(n)}$.
(iv) For any $\mathbf{a}\in R^n$ the formula (2.9) holds.

The class of elliptical distributions can be defined in a number of equivalent ways (see Kelker, 1970; Dempster, 1969; Eaton, 1981; Cambanis, Huang and Simons, 1981; Muirhead, 1982; Fang and Zhang, 1989).

Definition 2.2
An $n\times 1$ random vector $\mathbf{x}$ is said to have an **elliptically symmetric** distribution (or simply **elliptical** distribution) with parameters $\boldsymbol{\mu}(n\times 1)$ and $\mathbf{\Sigma}(n\times n)$ if

$$x\stackrel{d}{=}\boldsymbol{\mu}+\mathbf{A}'\mathbf{y}, \quad \mathbf{y}\sim S_k(\phi) \tag{2.10}$$

where $\mathbf{A}:k \times n$, $\mathbf{A}'\mathbf{A} = \mathbf{\Sigma}$ with $\operatorname{rank}(\mathbf{\Sigma}) = k$. We shall write $\mathbf{x} \sim EC_n(\boldsymbol{\mu}, \mathbf{\Sigma}, \phi)$.

The following two assertions can be easily verified and are left as an exercise to the reader. If $\mathbf{x} \sim EC_n(\boldsymbol{\mu}, \mathbf{\Sigma}, \phi)$ with $\operatorname{rank}(\mathbf{\Sigma}) = k$ then:

1. The c.f. of $\mathbf{x}$, $\psi(\mathbf{t}) = E(e^{i\mathbf{t}'\mathbf{x}})$ is of the form

$$\psi(\mathbf{t}) = e^{i\mathbf{t}'\boldsymbol{\mu}}\phi(\mathbf{t}'\mathbf{\Sigma}\mathbf{t}) \tag{2.11}$$

   for some scalar function ϕ.
2. $\mathbf{x}$ has a stochastic representation

$$\mathbf{x} \stackrel{d}{=} \boldsymbol{\mu} + r\mathbf{A}'\mathbf{u}^{(k)}, \tag{2.12}$$

   where $r \geqslant 0$ is independent of $\mathbf{u}^{(k)}$ and $\mathbf{A}'\mathbf{A} = \mathbf{\Sigma}$.

Some illustrative examples follow.

Example 2.4 (The multinormal distribution)
If random vector $\mathbf{x}$ has the following decomposition:

$$\mathbf{x} \stackrel{d}{=} \boldsymbol{\mu} + \mathbf{A}'\mathbf{y}$$

where $\boldsymbol{\mu} \in R^n$, $\mathbf{A}:m \times n$ and $\mathbf{y} \sim N_m(\mathbf{0}, \mathbf{I}_m)$, then we say that $\mathbf{x}$ has a multinormal distribution $N_n(\boldsymbol{\mu}, \mathbf{\Sigma})$ with $\mathbf{\Sigma} = \mathbf{A}'\mathbf{A}$. From Example 2.3, $\mathbf{y} \sim S_m(\phi)$ with $\phi(u) = \exp(-u/2)$, and we have that $\mathbf{x} \sim EC_n(\boldsymbol{\mu}, \mathbf{\Sigma}, \phi)$. Equivalently,

$$\mathbf{x} \stackrel{d}{=} \boldsymbol{\mu} + r\mathbf{A}'\mathbf{u}^{(k)} \tag{2.13}$$

with

$$r \stackrel{d}{=} \|\mathbf{y}\| \sim \chi_n \qquad \text{(chi-distribution with } n \text{ degrees of freedom).}$$

Example 2.5 (The multivariate t-distribution)
Let $\mathbf{z} \sim N_n(\mathbf{0}, \mathbf{I}_n)$ and $s \sim \chi_m$ be independent. Let

$$\mathbf{y} = m^{1/2}\mathbf{z}/s. \tag{2.14}$$

We say that $\mathbf{y}$ has a multivariate t-distribution with m degrees of freedom and write $\mathbf{y} \sim Mt_n(m, \mathbf{0}, \mathbf{I}_n)$.

Evidently, one can write (2.14) as follows:

$$\mathbf{y} \stackrel{d}{=} m^{1/2}r\mathbf{u}^{(n)}/s = r^*\mathbf{u}^{(n)}$$

where $r \sim \chi_n$, s and $\mathbf{u}^{(n)}$ are independent, and $r^* = m^{1/2}r/s$ (r^*/n has an F-distribution with n and m degrees of freedom). Thus $\mathbf{y}$ has a spherical distribution. Let

$$\mathbf{x} = \boldsymbol{\mu} + \mathbf{A}'\mathbf{y}, \qquad \mathbf{A}: n \times n \quad \text{and} \quad \boldsymbol{\mu} \in R^n.$$

We say that $\mathbf{x}$ has a multivariate t-distribution with parameters $\boldsymbol{\mu}$, $\boldsymbol{\Sigma} = \mathbf{A}'\mathbf{A}$, m degrees of freedom and write $\mathbf{x} \sim Mt_n(m, \boldsymbol{\mu}, \boldsymbol{\Sigma})$. Clearly, $Mt_n(m, \boldsymbol{\mu}, \boldsymbol{\Sigma}) = EC_n(\boldsymbol{\mu}, \boldsymbol{\Sigma}, \phi)$ with a special ϕ. More detail in Section 3.3.

2.2 Marginal distributions, moments and density

Many properties of spherical and elliptical distributions have been studied by Kelker (1970), Cambanis, Huang and Simons (1981), Anderson and Fang (1982), Dickey and Chen (1985, 1987), and Berkane and Bentler (1986). In this section we continue to discuss some of the properties of spherical distributions concentrating on marginal distributions, moments and density. In subsequent sections some further properties will be presented.

2.2.1 *Marginal distributions*

Let $\mathbf{x} = \begin{bmatrix} \mathbf{x}^{(1)} \\ \mathbf{x}^{(2)} \end{bmatrix} \sim S_n(\phi)$ where $\mathbf{x}^{(1)}: m \times 1$. It is obvious that $\mathbf{x}^{(1)} \sim S_m(\phi)$. That means that if $\mathbf{x} \sim S_n(\phi)$, then all the marginal distributions of $\mathbf{x}$ are spherical and all the marginal characteristic functions have the same generator. The following theorem is due to Anderson and Fang (1982).

Theorem 2.6

If $\mathbf{x} \sim S_n(\phi)$, partition $\mathbf{x}$ into m parts $\mathbf{x} = (\mathbf{x}^{(1)'}, \ldots, \mathbf{x}^{(m)'})'$ each with $n_1, \ldots, n_m$ components respectively. Then

$$\begin{bmatrix} \mathbf{x}^{(1)} \\ \vdots \\ \mathbf{x}^{(m)} \end{bmatrix} \overset{d}{=} \begin{bmatrix} rd_1\mathbf{u}_1 \\ \vdots \\ rd_m\mathbf{u}_m \end{bmatrix} \qquad (2.15)$$

where $d_i \geq 0$, $i = 1, \ldots, m$; $(d_1^2, \ldots, d_m^2) \sim D_m(n_1/2, \ldots, n_m/2)$; $\mathbf{u}_1, \ldots, \mathbf{u}_m$ and $(d_1^2, \ldots, d_m^2)$ are independent and $\mathbf{u}_j \overset{d}{=} \mathbf{u}^{(n_j)}$, $j = 1, \ldots, m$.

PROOF Let $\mathbf{y} \sim N_n(\mathbf{0}, \mathbf{I}_n)$ and $\mathbf{u}^{(n)}$ be partitioned in the same fashion as $\mathbf{x}$. By Theorem 2.3 we have

$$\mathbf{u}^{(n)}=\begin{bmatrix}\mathbf{u}^{(1)}\\ \vdots \\ \mathbf{u}^{(m)}\end{bmatrix}\stackrel{d}{=}\begin{bmatrix}\mathbf{y}^{(1)}/\|\mathbf{y}\|\\ \vdots \\ \mathbf{y}^{(m)}/\|\mathbf{y}\|\end{bmatrix}=\begin{bmatrix}\mathbf{y}^{(1)}/\|\mathbf{y}^{(1)}\|\cdot\|\mathbf{y}^{(1)}\|/\|\mathbf{y}\|\\ \vdots \\ \mathbf{y}^{(m)}/\|\mathbf{y}^{(m)}\|\cdot\|\mathbf{y}^{(m)}\|/\|\mathbf{y}\|\end{bmatrix}.$$

Let $d_j=\|\mathbf{y}^{(j)}\|/\|\mathbf{y}\|$ and $\mathbf{u}_j=\mathbf{y}^{(j)}/\|\mathbf{y}^{(j)}\|$, $j=1,\ldots,m$. Clearly, $(d_1,\ldots,d_m)$ satisfies the above required conditions in the statement of the theorem. Applying once more Theorem 2.3, we obtain $\mathbf{u}_j=\mathbf{y}^{(j)}/\|\mathbf{y}^{(j)}\|\stackrel{d}{=}\mathbf{u}^{(n_j)}$, $j=1,\ldots,m$. The theorem follows from the fact that $\{\mathbf{y}^{(j)},\ j=1,\ldots,m\}$ are independent, $\|\mathbf{y}^{(j)}\|$ is independent of $\mathbf{y}^{(j)}/\|\mathbf{y}^{(j)}\|$, and $\mathbf{x}\stackrel{d}{=}r\mathbf{u}^{(n)}$. □

2.2.2 *Moments*

Theorem 2.7

The mean vector and the covariance matrix of $\mathbf{u}^{(n)}$ are

$$E(\mathbf{u}^{(n)})=\mathbf{0},\qquad \mathrm{Cov}(\mathbf{u}^{(n)})=\frac{1}{n}\mathbf{I}_n, \tag{2.16}$$

respectively.

PROOF Let $\mathbf{x}\sim N_n(\mathbf{0},\mathbf{I}_n)$. From Theorem 2.3 we have $\mathbf{x}\stackrel{d}{=}\|\mathbf{x}\|\mathbf{u}^{(n)}$, where $\|\mathbf{x}\|$ is independent of $\mathbf{u}^{(n)}$. It is well known that $\|\mathbf{x}\|^2\sim\chi_n^2$. Since $E(\mathbf{x})=\mathbf{0}$, $E\|\mathbf{x}\|>0$, $E\|\mathbf{x}\|^2=n$ and $\mathrm{Cov}(\mathbf{x})=\mathbf{I}_n$, we arrive immediately at the assertion of the theorem. □

Corollary

If $\mathbf{x}=r\mathbf{u}^{(n)}\sim S_n(\phi)$, and $E(r^2)<\infty$, then

$$E(\mathbf{x})=\mathbf{0},\qquad \mathrm{Cov}(\mathbf{x})=\frac{E(r^2)}{n}\mathbf{I}_n. \tag{2.17}$$

Theorem 2.8

The moments of $\mathbf{x}=r\mathbf{u}^{(n)}$, provided they exist, can be expressed in terms of a one-dimensional integral. For even integers $2s_i\geqslant 0$, $i=1,\ldots,n$, we have

$$E\left(\prod_{i=1}^{n}X_i^{2s_i}\right)=E(r^{2s})\pi^{-n/2}\frac{\Gamma(n/2)}{\Gamma(n/2+s)}\prod_{i=1}^{n}\Gamma(\tfrac{1}{2}+s_i), \tag{2.18}$$

where $s=s_1+s_2+\cdots+s_n$.

This theorem is due to Dickey and Chen (1985). We shall prove it in Section 3.1 using another method.

2.2.3 *Density*

A random vector $\mathbf{x} \sim S_n(\phi)$, in general, does not necessarily possess a density. It is, however, easy to see from Theorem 2.1 that the density of $\mathbf{x}$, if it exists, must be of the form $g(\mathbf{x}'\mathbf{x})$ for some nonnegative function $g(\cdot)$ of a scalar variable. In this case, from (1.32) we have

$$\int g(\mathbf{x}'\mathbf{x})\,\mathrm{d}\mathbf{x} = \frac{\pi^{n/2}}{\Gamma(n/2)} \int_0^\infty y^{n/2-1} g(y)\,\mathrm{d}y = 1.$$

Hence, a nonnegative function $g(\cdot)$ can be used to define a density $cg(\mathbf{x}'\mathbf{x})$ for some spherical distribution if and only if

$$\int_0^\infty y^{n/2-1} g(y)\,\mathrm{d}y < \infty. \tag{2.19}$$

In this case, we shall write $\mathbf{x} \sim S_n(g)$ instead of $\mathbf{x} \sim S_n(\phi)$ and call $g(\cdot)$ a **density generator** of the spherical distribution, or simply the **p.d.f. generator**.

Theorem 2.9

If $\mathbf{x} \overset{d}{=} r\mathbf{u}^{(n)} \sim S_n(\phi)$, then $\mathbf{x}$ possesses a density generator $g(\cdot)$ if and only if r has a density $f(\cdot)$, and the relationship between $g(\cdot)$ and $f(\cdot)$ is as follows:

$$f(r) = \frac{2\pi^{n/2}}{\Gamma(n/2)} r^{n-1} g(r^2). \tag{2.20}$$

PROOF Assume that $\mathbf{x}$ possesses a density $g(\mathbf{x}'\mathbf{x})$. Let $h(\cdot)$ be any nonnegative measurable function. Using (1.32) We have

$$\begin{aligned} E[h(r)] &= \int h((\mathbf{x}'\mathbf{x})^{1/2}) g(\mathbf{x}'\mathbf{x})\,\mathrm{d}\mathbf{x} \\ &= \frac{\pi^{n/2}}{\Gamma(n/2)} \int_0^\infty h(y^{1/2}) y^{n/2-1} g(y)\,\mathrm{d}y \qquad (\text{letting } r = y^{1/2}) \\ &= \frac{2\pi^{n/2}}{\Gamma(n/2)} \int_0^\infty h(r) r^{n-1} g(r^2)\,\mathrm{d}r, \end{aligned}$$

which shows that r has the density $f(\cdot)$ of the form (2.20). The sufficiency is obvious. □

Remark: If $\mathbf{x} \sim S_n(g)$, then the density $h(u)$ of $u = r^2$ is

$$h(u) = \frac{\pi^{n/2}}{\Gamma(n/2)} u^{n/2-1} g(u) \tag{2.21}$$

Theorem 2.9 is due to Kelker (1970) and the proof is adapted from Fang and Zhang (1989).

Theorem 2.10

Assume that $\mathbf{x} \stackrel{d}{=} r\mathbf{u}^{(n)} \sim S_n^+(\phi)$. Then all the marginal distributions of $\mathbf{x}$ possess densities. In particular, the marginal density of $(x_1, \ldots, x_k)$ $(1 \leqslant k < n)$ is

$$\frac{\Gamma(n/2)}{\Gamma((n-k)/2)\pi^{k/2}} \int_{(\sum_{i=1}^k x_i^2)^{1/2}}^{\infty} r^{-(n-2)} \left(r^2 - \sum_{i=1}^k x_i^2 \right)^{(n-k)/2-1} \mathrm{d}F(r), \tag{2.22}$$

where $F(r)$ is the c.d.f. of r.

PROOF We shall prove (2.22) only. The marginal density of $(u_1, \ldots, u_k)$ of $\mathbf{u}^{(n)}$ is (cf. (1.26))

$$\frac{\Gamma(n/2)}{\Gamma((n-k)/2)\pi^{k/2}} \left(1 - \sum_{i=1}^k u_i^2 \right)^{(n-k)/2-1}, \qquad \left(\sum_{i=1}^k u_i^2 < 1 \right).$$

Hence, the c.d.f. of $(X_1, \ldots, X_k)$ is

$$\frac{\Gamma(n/2)}{\Gamma((n-k)/2)\pi^{k/2}} \int_0^{\infty} \mathrm{d}F(r) \int_{-1}^{x_1/r} \cdots \int_{-1}^{x_k/r} \left(1 - \sum_{i=1}^k u_i^2 \right)^{(n-k)/2-1}$$
$$\times I(u_1, \ldots, u_k) \, \mathrm{d}u_1 \cdots \mathrm{d}u_k,$$

where

$$I(u_1, \ldots, u_k) = \begin{cases} 1, & \sum_{i=1}^k u_i^2 < 1 \\ 0, & \text{otherwise} \end{cases}$$

Differentiating the above formula with respect to $x_1, \ldots, x_k$, we arrive at (2.22). □

If $\mathbf{x} \sim S_n(g)$, then all the marginal densities exist. Denote the p.d.f. of $(x_1, \ldots, x_m)$ by $g_m(x_1^2 + \cdots + x_m^2)$, $1 \leqslant m \leqslant n$. By using Lemma 1.3 with $\alpha_i = 1$, $i = 1, \ldots, m$, we can find

$$g_m(u) = \frac{\pi^{(n-m)/2}}{\Gamma((n-m)/2)} \int_u^{\infty} (y-u)^{(n-m)/2-1} g(y) \, \mathrm{d}(y). \tag{2.23}$$

In particular,

$$g_m(u) = \pi \int_u^{\infty} g_{m+2}(y)\,dy.$$

From the above formulas we immediately arrive at the following results:

1. If $\mathbf{x} \sim S_n(g)$, the marginal densities of dimensions less than or equal to $n-1$ are continuous and the marginal densities of dimensions less than or equal to $n-2$ are differentiable (except possibly at the origin in both cases).
2. Univariate marginal densities for $n \geqslant 2$ are nondecreasing on $(-\infty, 0)$ and nonincreasing on $(0, \infty)$.
3. For $1 \leqslant m \leqslant n-2$, the (marginal) density generators are related by

$$g_{m+2}(x) = (-1/\pi)g'_m(x) \quad \text{for } x > 0 \quad \text{(a.e.)}, \tag{2.24}$$

   where $g'_m(\ )$ is the derivative of $g_m(\ )$.
4. Equation (2.24) enables one to construct all of the marginal densities if only the univariate marginal density is known.

Theorem 2.10 is adapted from Kelker (1970) and Anderson and Fang (1982).

Theorem 2.11

Suppose $\mathbf{x}' = (x_1, \ldots, x_n) \sim S_n(g)$, $n \geqslant 2$. Consider the transformation to spherical coordinates for $\mathbf{x}$,

$$\begin{cases} x_j = r\left(\prod_{k=1}^{j-1} \sin\theta_k\right)\cos\theta_j, & 1 \leqslant j \leqslant n-1 \\ x_n = r\left(\prod_{k=1}^{n-2} \sin\theta_k\right)\sin\theta_{n-1}, & \end{cases} \tag{2.25}$$

where $r \geqslant 0$, $\theta_k \in [0, \pi)$, $k = 1, \ldots, n-2$, $\theta_{n-1} \in [0, 2\pi)$. Then r, $\theta_1, \ldots, \theta_{n-1}$ are independent and have respectively the p.d.f.s

$$\begin{cases} h_r(r) = \dfrac{2\pi^{n/2}}{\Gamma(n/2)} r^{n-1} g(r^2), & r \geqslant 0, \\ h_{\theta_k}(\theta_k) = \dfrac{1}{B(\frac{1}{2}, (n-k)/2)} \sin^{n-k-1}\theta_k, & 0 \leqslant \theta_k < \pi, k = 1, \ldots, n-2, \\ h_{\theta_{n-1}}(\theta_{n-1}) = 1/2\pi, & 0 \leqslant \theta_{n-1} < 2\pi. \end{cases} \tag{2.26}$$

Conversely if $r, \theta_1, \ldots, \theta_{n-1}$ are independent and have p.d.f.s given by (2.26), and $\mathbf{x}$ is defined by (2.25), then $\mathbf{x} \sim S_n(g)$.

PROOF The density of $\mathbf{x}$ is $g(\mathbf{x}'\mathbf{x})$. For the transformation

$$x_j = r\left(\prod_{k=1}^{j-1} \sin\theta_k\right)\cos\theta_j, \qquad 1 \leqslant j \leqslant n-1$$

$$x_n = r\left(\prod_{k=1}^{n-2} \sin\theta_k\right)\sin\theta_{n-1},$$

the Jacobian is $r^{n-1}\prod_{k=1}^{n-2}\sin^{n-k-1}\theta_k$ (e.g. Graham, 1981). Hence the joint density of $(r, \theta_1, \ldots, \theta_{n-1})$ is

$$g(r^2)r^{n-1}\prod_{k=1}^{n-2}\sin^{n-k-1}\theta_k,$$

which implies (2.26) and the independence of $r, \theta_1, \ldots,$ and θ_{n-1}. □

This theorem is due to Goldman (1974, 1976), Muirhead (1980) – see also Cacoullos and Koutras (1984).

2.3 The relationship between ϕ and F

As we have seen in the preceding sections, a spherical distribution can be obtained via its characteristic generator ϕ with the c.f. $\phi(\mathbf{t}'\mathbf{t})$ or by its generating variate r with the c.d.f. $F(\cdot)$ and $r\mathbf{u}^{(n)}$. The formula (2.5) (Theorem 2.2) gives the relationship between ϕ and F. The following theorem will show that the relationship (2.5) between $\phi \in \Phi_n$ and F is one-to-one.

Theorem 2.12
Assume $\mathbf{x} \overset{d}{=} r_1\mathbf{u}^{(n)} \overset{d}{=} r_2\mathbf{u}^{(n)} \sim S_n(\phi)$. Then $r_1 \overset{d}{=} r_2$.

PROOF Since $\mathbf{x}'\mathbf{x} \overset{d}{=} r_1^2 \overset{d}{=} r_2^2$, we have $r_1 \overset{d}{=} r_2$. □

The theorem implies that 'the number' of spherical distributions is essentially equal to 'the number' of nonnegative random variables.

Let $\phi(\cdot) \in \Phi_n$, then for each $1 \leqslant m \leqslant n$, $\phi \in \Phi_m$. Because of the relationship (2.5) between ϕ and F, there exist two c.d.f.s F_n and F_m

such that

$$\phi(x) = \int_0^\infty \Omega_n(xu^2)\,\mathrm{d}F_n(u)$$

and

$$\phi(x) = \int_0^\infty \Omega_m(xu^2)\,\mathrm{d}F_m(u).$$

The precise relationship between the various F_n and F_m is specified in the following result.

Corollary
If r_n and r_m possess c.d.f.s F_n and F_m, respectively, then $r_m \stackrel{d}{=} r_n b$, where $b \geqslant 0$ and $b^2 \sim Be(m/2, (n-m)/2)$ is independent of r_n.

PROOF Let $\mathbf{x} = (\mathbf{x}^{(1)'}, \mathbf{x}^{(2)'})' \sim S_n(\phi)$, where $\mathbf{x}^{(1)}$:$m \times 1$, then the c.f. of $\mathbf{x}^{(1)}$ is $\phi(\|\mathbf{t}^{(1)}\|^2)$, where $\mathbf{t}^{(1)}$:$m \times 1$. On the other hand, by Theorem 2.6, $\mathbf{x}^{(1)} \stackrel{d}{=} r_n d_1 \mathbf{u}^{(m)}$, completes the proof. □

The above relationship between ϕ and r_n (or r_m) can be written as $\phi \in \Phi_n \leftrightarrow r_n$ (or F_n) and $\phi \in \Phi_m \leftrightarrow r_m \stackrel{d}{=} r_n B_{m/2,(n-m)/2}$ (or F_m). It can be verified by a direct argument that r_m has a density over $(0, \infty)$, given by

$$f_m(x) = \frac{2x^{m-1}\Gamma(\frac{1}{2}n)}{\Gamma(\frac{1}{2}m)\Gamma(\frac{1}{2}(n-m))} \int_x^\infty r^{-(n-2)}(r^2 - x^2)^{(n-m)/2-1}\,\mathrm{d}F_n(r). \tag{2.27}$$

In Section 2.5 we shall pay special attention to the class Φ_∞.

2.4 Conditional distributions

Let $\mathbf{x}$ be $n \times 1$ random and $\mathbf{x} \sim (\mathbf{x}^{(1)'}, \mathbf{x}^{(2)'})'$, where $\mathbf{x}^{(1)}$:$m \times 1$, $0 < m < n$. We shall now consider the conditional distribution of $\mathbf{x}^{(1)}$ given $\mathbf{x}^{(2)}$. When $\mathbf{x}$ has p.d.f. $g(\mathbf{x}'\mathbf{x})$, $\mathbf{x}^{(2)}$ also has a p.d.f. of the form $g_2(\mathbf{x}^{(2)'}\mathbf{x}^{(2)})$. The conditional p.d.f. of $\mathbf{x}^{(1)}$ given $\mathbf{x}^{(2)}$ is then

$$\begin{aligned} h(\mathbf{x}^{(1)}|\mathbf{x}^{(2)}) &= g(\mathbf{x}'\mathbf{x})/g_2(\mathbf{x}^{(2)'}\mathbf{x}^{(2)}) \\ &= g(\mathbf{x}^{(1)'}\mathbf{x}^{(1)} + \mathbf{x}^{(2)'}\mathbf{x}^{(2)})/g_2(\mathbf{x}^{(2)'}\mathbf{x}^{(2)}). \end{aligned}$$

This means that $h(\mathbf{x}^{(1)}|\mathbf{x}^{(2)})$ is a spherical density in R^m. Problem 2.3

shows that $(\mathbf{x}^{(1)}|\mathbf{x}^{(2)}) \sim S_m(g_a)$ with

$$g_a(y) = \frac{\Gamma(m/2)g(a+y)}{\pi^{m/2}\int_0^\infty u^{m/2-1}g(a+u)\,du},$$

where $a = (\mathbf{x}^{(2)})'(\mathbf{x}^{(2)})$.

The following theorem shows that even when $\mathbf{x}$ does not possess a p.d.f. the conditional distribution of $\mathbf{x}^{(1)}$ given $\mathbf{x}^{(2)}$ is still spherical.

Theorem 2.13

Suppose $\mathbf{x} \stackrel{d}{=} r\mathbf{u}^{(n)} \sim S_n(\phi)$, then the conditional distribution of $\mathbf{x}^{(1)}$ given $\mathbf{x}^{(2)} = \mathbf{x}_0^{(2)}$ is given by

$$(\mathbf{x}^{(1)}|\mathbf{x}^{(2)} = \mathbf{x}_0^{(2)}) \sim S_m(\phi_{\|\mathbf{x}_0^{(2)}\|^2}), \tag{2.28}$$

with the stochastic representation

$$(\mathbf{x}^{(1)}|\mathbf{x}^{(2)} = \mathbf{x}_0^{(2)}) \stackrel{d}{=} r(\|\mathbf{x}_0^{(2)}\|^2)\mathbf{u}^{(m)} \tag{2.29}$$

where for each $a^2 \geqslant 0$, $r(a^2)$ and $\mathbf{u}^{(m)}$ are independent and

$$r(\|\mathbf{x}_0^{(2)}\|^2) \stackrel{d}{=} ((r^2 - \|\mathbf{x}_0^{(2)}\|^2)^{1/2}|\mathbf{x}^{(2)} = \mathbf{x}_0^{(2)}) \tag{2.30}$$

and the function ϕ_{a^2} is given by (2.5) with $n = m$ and F replaced by the distribution function of $r(a^2)$.

PROOF From Theorem 2.6, we have

$$\begin{bmatrix} \mathbf{x}^{(1)} \\ \mathbf{x}^{(2)} \end{bmatrix} \stackrel{d}{=} \begin{bmatrix} rd_1\mathbf{u}^{(m)} \\ rd_2\mathbf{u}^{(n-m)} \end{bmatrix},$$

where r, d_1 (or d_2), $\mathbf{u}^{(m)}$ and $\mathbf{u}^{(n-m)}$ are independent, $d_1^2 + d_2^2 = 1$, and $d_1^2 \sim \mathrm{Be}(m/2, (n-m)/2)$. For $rd_2\mathbf{u}^{(n-m)} = \mathbf{x}_0^{(2)}$, it is easy to verify that $rd_1 = (r^2 - \|\mathbf{x}_0^{(2)}\|^2)^{1/2}$ and (2.30) and (2.29) follow. That $r(\|\mathbf{x}_0^{(2)}\|^2)$, as defined in (2.30), depends on $\mathbf{x}_0^{(2)}$ only through $\|\mathbf{x}_0^{(2)}\|^2$, is obvious when $\mathbf{x}_0^{(2)} = \mathbf{0}$, while for $\mathbf{x}_0^{(2)} \neq \mathbf{0}$, we have

$$\begin{aligned} ((r^2 - \|\mathbf{x}_0^{(2)}\|^2)^{1/2}|\mathbf{x}^{(2)} = \mathbf{x}_0^{(2)}) &\stackrel{d}{=} ((r^2 - \|\mathbf{x}_0^{(2)}\|^2)^{1/2}|rd_2\mathbf{u}^{(n-m)} = \mathbf{x}_0^{(2)}) \\ &\stackrel{d}{=} ((r^2 - \|\mathbf{x}_0^{(2)}\|^2)^{1/2}|r^2d_2^2 = \|\mathbf{x}_0^{(2)}\|^2), \end{aligned}$$

since $\mathbf{u}^{(n-m)}$ is independent of r and d_1. Thus (2.28) follows from

$$(\mathbf{x}^{(1)}|\mathbf{x}^{(2)}=\mathbf{x}_0^{(2)}) \stackrel{d}{=} (rd_1\mathbf{u}^{(m)}|\mathbf{x}^{(2)}=\mathbf{x}_0^{(2)})$$

$$\stackrel{d}{=} ((r^2-\|\mathbf{x}_0^{(2)}\|^2)^{1/2}\mathbf{u}^{(m)}|\mathbf{x}^{(2)}=\mathbf{x}_0^{(2)})$$

$$\sim S_m(\phi_{\|\mathbf{x}_0^{(2)}\|^2}). \quad \square$$

Corollary
In the notation of Theorem 2.13 we have

$$r(a^2)=0 \text{ a.s.} \quad \text{for } a=0 \quad \text{or} \quad F(a)=1,$$

and

$$P(r(a^2)\leqslant\rho)=\frac{\int_a^{(a^2+\rho^2)^{1/2}}(r^2-a^2)^{m/2-1}r^{-(n-2)}\,\mathrm{d}F(r)}{\int_a^{\infty}(r^2-a^2)^{m/2-1}r^{-(n-2)}\,\mathrm{d}F(r)} \tag{2.31}$$

for $\rho\geqslant 0$, $a>0$ and $F(a)<1$.

This corollary is based on the following lemma, whose proof is based on straightforward results and is therefore omitted.

Lemma 2.1
Suppose that the nonnegative random variables R and S are independent, R has distribution function F, and S has a density g. Then $T=RS$ has an atom of size $F(0)$ at zero; if $F(0)>0$, it is absolutely continuous on $(0,\infty)$ with density h given by

$$h(t)=\int_{(0,\infty)} r^{-1}g(t/r)\,\mathrm{d}F(r),$$

and a regular version of the conditional distribution of R given $T=t$ is expressible in the form

$$P(R\leqslant r|T=t)=\begin{cases}0 & \text{for } r<0;\\ 1 & \text{for } t=0 \quad \text{or} \quad t>0 \quad \text{with } h(t)=0,\ r\geqslant 0;\\ (h(t))^{-1}\displaystyle\int_{(0,r]} x^{-1}g(t/x)\,\mathrm{d}F(x) & \text{for } t>0 \text{ with } h(t)\neq 0,\ r\geqslant 0.\end{cases}$$

The material of this section is from Cambanis, Huang and Simons (1981). These authors further point out that the distribution function

F_{a^2} of $r(a^2)$ is determined by F in the following manner.

$$F(r) = 1 - C_{a^2} \int_{(r^2 - a^2)^{1/2}} (\rho^2 + a^2)^{n/2 - 1} \rho^{-(m-2)} \mathrm{d}F_{a^2}(\rho),$$

$$r \geqslant \rho, \quad a > 0, \quad (2.32)$$

where C_{a^2} is the denominator in (2.31). Thus, whenever $a > 0$, F_{a^2} determines F on the interval $[a, \infty)$ up to the unknown multiplicative factor $C_{a^2} \geqslant 0$, and, of course, contains no information about the values of F on $[0, a)$. If $F(r)$ is known for some $r \geqslant a$ and $F(r) < 1$, then F_{a^2} determines F uniquely on $[a, \infty)$ by means of (2.32).

2.5 Properties of elliptically symmetric distributions

In the last four sections, we have discussed the properties of spherical distributions. In this section, we shall proceed to an analogous discussion of elliptically symmetric distributions defined in Section 2.1. Many proofs will be omitted. More details are given in Cambanis, Huang and Simons (1981) and Chu (1973).

Theorem 2.14
A scalar function $\phi(\cdot)$ can determine an elliptically symmetric distribution $EC_n(\boldsymbol{\mu}, \boldsymbol{\Sigma}, \phi)$ for every $\boldsymbol{\mu} \in R^n$ and $\boldsymbol{\Sigma} \geqslant 0$ with $\operatorname{rank}(\boldsymbol{\Sigma}) = k$ if and only if $\phi \in \Phi_k$.

Corollary 1
$\mathbf{x} \sim EC_n(\boldsymbol{\mu}, \boldsymbol{\Sigma}, \phi)$ with $\operatorname{rank}(\boldsymbol{\Sigma}) = k$ if and only if

$$\mathbf{x} \stackrel{d}{=} \boldsymbol{\mu} + r\mathbf{A}'\mathbf{u}^{(k)}, \tag{2.33}$$

where $r \geqslant 0$ is independent of $\mathbf{u}^{(k)}$, $r \leftrightarrow \phi \in \Phi_k$, and $\mathbf{A}$ is a $k \times n$ matrix such that $\mathbf{A}'\mathbf{A} = \boldsymbol{\Sigma}$.

Corollary 2
Assume that $\mathbf{x} \sim EC_n(\boldsymbol{\mu}, \boldsymbol{\Sigma}, \phi)$ with $\operatorname{rank}(\boldsymbol{\Sigma}) = k$. Then

$$Q(\mathbf{x}) = (\mathbf{x} - \boldsymbol{\mu})'\boldsymbol{\Sigma}^{-}(\mathbf{x} - \boldsymbol{\mu}) \stackrel{d}{=} r^2, \tag{2.34}$$

where $\boldsymbol{\Sigma}^{-}$ is a generalized inverse of $\boldsymbol{\Sigma}$ (cf. Section 1.2).

Theorem 2.15
Assume that $\mathbf{x}$ is nondegenerate.

(i) If $\mathbf{x} \sim EC_n(\boldsymbol{\mu}, \boldsymbol{\Sigma}, \phi)$ and $\mathbf{x} \sim EC_n(\boldsymbol{\mu}^*, \boldsymbol{\Sigma}^*, \phi^*)$, then there exists a constant $c > 0$, such that

$$\boldsymbol{\mu}^* = \boldsymbol{\mu}, \quad \boldsymbol{\Sigma}^* = c\boldsymbol{\Sigma}, \quad \phi^*(\cdot) = \phi(c^{-1}\cdot). \tag{2.35}$$

(ii) If $\mathbf{x} \stackrel{d}{=} \boldsymbol{\mu} + r\mathbf{A}'\boldsymbol{\mu}^{(l)} \stackrel{d}{=} \boldsymbol{\mu}^* + r^*\mathbf{A}^{*\prime}\mathbf{u}^{(l^*)}$, where $l \geqslant l^*$, then there exists a constant $c > 0$ such that

$$\boldsymbol{\mu}^* = \boldsymbol{\mu}, \quad \mathbf{A}^{*\prime}\mathbf{A}^* = c\mathbf{A}'\mathbf{A}, \quad r^* \stackrel{d}{=} c^{-1/2}rb, \tag{2.36}$$

where $b \geqslant 0$ is independent of r and $b^2 \sim \mathrm{Be}(\frac{1}{2}l^*, \frac{1}{2}(l - l^*))$ if $l > l^*$, and $b \equiv 1$ if $l = l^*$.

This theorem shows that $\boldsymbol{\Sigma}$, ϕ, r, $\mathbf{A}$ are not unique unless we impose the condition that $|\boldsymbol{\Sigma}| = 1$ or that $|\mathbf{A}'\mathbf{A}| = 1$. The next theorem points out that any linear combination of elliptically distributed variates is still elliptical.

Theorem 2.16
Assume that $\mathbf{x} \sim EC_n(\boldsymbol{\mu}, \boldsymbol{\Sigma}, \phi)$ with $\mathrm{rank}(\boldsymbol{\Sigma}) = k$, $\mathbf{B}$ is an $n \times m$ matrix and $\boldsymbol{\nu}$ is an $m \times 1$ vector, then

$$\boldsymbol{\nu} + \mathbf{B}'\mathbf{x} \sim EC_m(\boldsymbol{\nu} + \mathbf{B}'\boldsymbol{\mu}, \mathbf{B}'\boldsymbol{\Sigma}\mathbf{B}, \phi) \tag{2.37}$$

PROOF The theorem follows directly from

$$\boldsymbol{\nu} + \mathbf{B}'\mathbf{x} \stackrel{d}{=} (\boldsymbol{\nu} + \mathbf{B}'\boldsymbol{\mu}) + r(\mathbf{A}\mathbf{B})'\mathbf{u}^{(k)}. \quad \square$$

Partition $\mathbf{x}$, $\boldsymbol{\mu}$, $\boldsymbol{\Sigma}$ into

$$\mathbf{x} = \begin{pmatrix} \mathbf{x}^{(1)} \\ \mathbf{x}^{(2)} \end{pmatrix} \quad \boldsymbol{\mu} = \begin{pmatrix} \boldsymbol{\mu}^{(1)} \\ \boldsymbol{\mu}^{(2)} \end{pmatrix} \quad \boldsymbol{\Sigma} = \begin{pmatrix} \boldsymbol{\Sigma}_{11} & \boldsymbol{\Sigma}_{12} \\ \boldsymbol{\Sigma}_{21} & \boldsymbol{\Sigma}_{22} \end{pmatrix} \tag{2.38}$$

where $\mathbf{x}^{(1)}$: $m \times 1$, $\boldsymbol{\mu}^{(1)}$: $m \times 1$ and $\boldsymbol{\Sigma}_{11}$: $m \times m$. The following corollary provides the marginal distributions which are also elliptical.

Corollary
Assume that $\mathbf{x} \sim EC_n(\boldsymbol{\mu}, \boldsymbol{\Sigma}, \phi)$, then $\mathbf{x}^{(1)} \sim EC_m(\boldsymbol{\mu}^{(1)}, \boldsymbol{\Sigma}_{11}, \phi)$ and $\mathbf{x}^{(2)} \sim EC_{n-m}(\boldsymbol{\mu}^{(2)}, \boldsymbol{\Sigma}_{22}, \phi)$.

Theorem 2.17
Assume that $\mathbf{x} \sim EC_n(\boldsymbol{\mu}, \boldsymbol{\Sigma}, \phi)$ and $E(r^2) < \infty$. Then

$$E(\mathbf{x}) = \boldsymbol{\mu}, \quad \mathrm{Cov}(\mathbf{x}) = \frac{E(r^2)}{\mathrm{rank}(\boldsymbol{\Sigma})}\boldsymbol{\Sigma} = -2\phi'(0)\boldsymbol{\Sigma}, \tag{2.39}$$

$$\Gamma_2(\mathbf{x}) = E(\mathbf{x}\mathbf{x}') = \boldsymbol{\mu}\boldsymbol{\mu}' - 2\phi'(0)\boldsymbol{\Sigma} \tag{2.40}$$

where $\phi'(0)$ is the derivative of ϕ at the origin and

$$\Gamma_1(\mathbf{x}) = i^{-1}\frac{\partial\phi(\mathbf{t})}{\partial\mathbf{t}}\bigg|_{\mathbf{t}=\mathbf{0}}$$

$$\Gamma_2(\mathbf{x}) = i^{-2}\frac{\partial^2\phi(\mathbf{t})}{\partial\mathbf{t}\,\partial\mathbf{t}'}\bigg|_{\mathbf{t}=\mathbf{0}}$$

PROOF Denoting $k = \text{rank}(\boldsymbol{\Sigma})$, we have

$$\mathbf{x} \stackrel{d}{=} \boldsymbol{\mu} + r\mathbf{A}'\mathbf{u}^{(k)}.$$

By Theorem 2.7, we obtain

$$E(\mathbf{x}) = \boldsymbol{\mu} + E(r)\mathbf{A}'E(\mathbf{u}^{(k)}) = \boldsymbol{\mu},$$

and

$$\begin{aligned}\text{Cov}(\mathbf{x}) &= \text{Cov}(r\mathbf{A}'\mathbf{u}^{(k)}) = E(r^2)\mathbf{A}'\text{Cov}(\mathbf{u}^{(k)})\mathbf{A}\\ &= E(r^2)\frac{1}{k}\mathbf{A}'\mathbf{I}_k\mathbf{A} = \frac{1}{k}E(r^2)\cdot\boldsymbol{\Sigma},\end{aligned}$$

which completes the proof of the first part of (2.39). The remaining assertions follow directly from straightforward calculations. The details are thus omitted (cf. Li, 1984). □

For expressions of moments and cumulants of higher order and Edgeworth expansion, see a recent paper by Traat (1989). Clévonx and Ducharme (1989) have considered a measure of multivariate correlation and its asymptotic distribution.

Muirhead (1982) defines the kurtosis parameter κ of an elliptical distribution $EC_n(\boldsymbol{\mu}, \boldsymbol{\Sigma}, \phi)$ as

$$\kappa = \frac{\phi''(0)}{(\phi'(0))^2} - 1,$$

where $\phi''(0)$ and $\phi'(0)$ are the second and first derivatives of ϕ, evaluated at zero. Berkane and Bentler (1986) define the general moment parameters as

$$\kappa(m) = \frac{\phi^{(m)}(0)}{(\phi'(0))^m} - 1,$$

and show that for $\mathbf{X} \sim EC_n(\mathbf{0}, \boldsymbol{\Sigma}, \phi)$, the moment $\mu_{i_1,i_2,\ldots,i_{2m}} = E(X_{i_1}X_{i_2}\cdots X_{i_{2m}})$ can be computed by

$$\mu_{i_1,i_2,\ldots,i_{2m}} = (\kappa(m)+1)\Sigma\mu_{i_1,i_2}\mu_{i_3,i_4}\cdots\mu_{i_{2m-1},i_{2m}},$$

where the summation is over all the $(2m)!/(2^m m!)$ possible groupings of the subscripts into pairs. For example,

$$\mu_{ijkl} = (\kappa + 1)(\mu_{ij}\mu_{kl} + \mu_{ik}\mu_{jl} + \mu_{il}\mu_{jk}).$$

For an n-dimensional random vector $\mathbf{x}$ with $E(\mathbf{x}) = \boldsymbol{\mu}$ and $\mathrm{Cov}(\mathbf{x}) = \boldsymbol{\Sigma}$, Mardia (1970) defined the **coefficient of multivariate kurtosis** as

$$\kappa_1 = E((\mathbf{x} - \boldsymbol{\mu})'\boldsymbol{\Sigma}^{-1}(\mathbf{x} - \boldsymbol{\mu}))^2/n(n+2) - 1.$$

Berkane and Bentler (1987) pointed out that $\kappa_1 = \kappa$ provided that $\mathbf{x} \sim EC_n(\boldsymbol{\mu}, \boldsymbol{\Sigma}, \phi)$ and $\mathrm{Cov}(\mathbf{x}) = \boldsymbol{\Sigma}$.

Theorem 2.18

Let $\mathbf{x} \overset{d}{=} \boldsymbol{\mu} + r\mathbf{A}'\mathbf{u}^{(n)} \sim EC_n(\boldsymbol{\mu}, \boldsymbol{\Sigma}, \phi)$ with $\boldsymbol{\Sigma} > 0$. Partition $\mathbf{x}$, $\boldsymbol{\mu}$, $\boldsymbol{\Sigma}$ into

$$\mathbf{x} = \begin{pmatrix} \mathbf{x}^{(1)} \\ \mathbf{x}^{(2)} \end{pmatrix}, \qquad \boldsymbol{\mu} = \begin{pmatrix} \boldsymbol{\mu}^{(1)} \\ \boldsymbol{\mu}^{(2)} \end{pmatrix}, \qquad \boldsymbol{\Sigma} = \begin{pmatrix} \boldsymbol{\Sigma}_{11} & \boldsymbol{\Sigma}_{12} \\ \boldsymbol{\Sigma}_{21} & \boldsymbol{\Sigma}_{22} \end{pmatrix},$$

where $\mathbf{x}^{(1)}$: $m \times 1$, $\boldsymbol{\mu}^{(1)}$: $m \times 1$, $\boldsymbol{\Sigma}_{11}$: $m \times m$ and $0 < m < n$. Then

$$(\mathbf{x}^{(1)} | \mathbf{x}^{(2)} = \mathbf{x}_0^{(2)}) \overset{d}{=} \boldsymbol{\mu}_{1.2} + r_{q(\mathbf{x}_0^{(2)})}\mathbf{A}'_{11.2}\mathbf{u}^{(n)}$$

$$\sim EC_m(\boldsymbol{\mu}_{1.2}, \boldsymbol{\Sigma}_{11.2}, \phi_{q(\mathbf{x}_0^{(2)})}) \tag{2.41}$$

where

$$\begin{cases} \boldsymbol{\mu}_{1.2} = \boldsymbol{\mu}^{(1)} + \boldsymbol{\Sigma}_{12}\boldsymbol{\Sigma}_{22}^{-1}(\mathbf{x}_0^{(2)} - \boldsymbol{\mu}^{(2)}), \\ \boldsymbol{\Sigma}_{11.2} = \boldsymbol{\Sigma}_{11} - \boldsymbol{\Sigma}_{12}\boldsymbol{\Sigma}_{22}^{-1}\boldsymbol{\Sigma}_{21} = \mathbf{A}'_{11.2}\mathbf{A}_{11.2}, \\ q(\mathbf{x}_0^{(2)}) = (\mathbf{x}_0^{(2)} - \boldsymbol{\mu}^{(2)})'\boldsymbol{\Sigma}_{22}^{-1}(\mathbf{x}_0^{(2)} - \boldsymbol{\mu}^{(2)}). \end{cases} \tag{2.42}$$

Moreover, for each $a \geqslant 0$, $r(a^2)$ is independent of $\mathbf{u}^{(m)}$ and its distribution is given by (2.30),

$$r_{q(\mathbf{x}_0^{(2)})} \overset{d}{=} ((r^2 - q(\mathbf{x}_0^{(2)}))^{1/2} | \mathbf{x}^{(2)} = \mathbf{x}_0^{(2)}),$$

the function ϕ_{a^2} is given by (2.5) with $n = m$ and F being replaced by the c.d.f. of $r(a^2)$.

PROOF Let $\mathbf{A}_2$ be a matrix such that $\mathbf{A}'_2\mathbf{A}_2 = \boldsymbol{\Sigma}_{22}$, $\mathbf{A}_2 > \mathbf{0}$ and

$$\boldsymbol{\Sigma} = \mathbf{A}'\mathbf{A} = \begin{pmatrix} \mathbf{I} & \boldsymbol{\Sigma}_{12}\boldsymbol{\Sigma}_{22}^{-1} \\ \mathbf{0} & \mathbf{I} \end{pmatrix} \begin{pmatrix} \boldsymbol{\Sigma}_{11.2} & \mathbf{0} \\ \mathbf{0} & \boldsymbol{\Sigma}_{22} \end{pmatrix} \begin{pmatrix} \mathbf{I} & \mathbf{0} \\ \boldsymbol{\Sigma}_{22}^{-1}\boldsymbol{\Sigma}_{21} & \mathbf{I} \end{pmatrix} = \mathbf{F}'\mathbf{F}$$

where

$$\mathbf{F} = \begin{pmatrix} \mathbf{A}_{11.2} & \mathbf{0} \\ \mathbf{0} & \mathbf{A}_2 \end{pmatrix} \begin{pmatrix} \mathbf{I} & \mathbf{0} \\ \boldsymbol{\Sigma}_{22}^{-1}\boldsymbol{\Sigma}_{21} & \mathbf{I} \end{pmatrix} = \begin{pmatrix} \mathbf{A}_{11.2} & \mathbf{0} \\ \mathbf{A}_2\boldsymbol{\Sigma}_{22}^{-1}\boldsymbol{\Sigma}_{21} & \mathbf{A}_2 \end{pmatrix},$$

$\boldsymbol{\Sigma}_{11.2} = \mathbf{A}'_{11.2}\mathbf{A}_{11.2}$.

Let $\mathbf{y} \sim EC_n(\mathbf{0}, \mathbf{I}_n, \phi)$, then we have

$$\mathbf{x} \overset{d}{=} \boldsymbol{\mu} + \mathbf{A}'\mathbf{y} \overset{d}{=} \boldsymbol{\mu} + \mathbf{F}'\mathbf{y} = \begin{pmatrix} \boldsymbol{\mu}^{(1)} \\ \boldsymbol{\mu}^{(2)} \end{pmatrix} + \begin{pmatrix} \mathbf{A}'_{11.2}\mathbf{y}^{(1)} + \boldsymbol{\Sigma}_{12}\boldsymbol{\Sigma}_{22}^{-1}\mathbf{A}'_2\mathbf{y}^{(2)} \\ \mathbf{A}'_2\mathbf{y}^{(2)} \end{pmatrix}$$

and

$$\begin{aligned}(\mathbf{x}^{(1)}|\mathbf{x}^{(2)}) &= (\mathbf{A}'_{11.2}\mathbf{y}^{(1)} + \boldsymbol{\Sigma}_{12}\boldsymbol{\Sigma}_{22}^{-1}\mathbf{A}'_2\mathbf{y}^{(2)} + \boldsymbol{\mu}^{(1)}|\boldsymbol{\mu}^{(2)} + \mathbf{A}_2\mathbf{y}^{(2)} = \mathbf{x}^{(2)}) \\ &= (\boldsymbol{\mu}^{(1)} + \boldsymbol{\Sigma}_{12}\boldsymbol{\Sigma}_{22}^{-1}(\mathbf{x}^{(2)} - \boldsymbol{\mu}^{(2)}) + \mathbf{A}'_{11.2}\mathbf{y}^{(1)} \\ &\quad |\mathbf{y}^{(2)} = \mathbf{A}_2^{-1}(\mathbf{x}^{(2)} - \boldsymbol{\mu}^{(2)})).\end{aligned}$$

From the above the result follows as a consequence of Theorem 2.13. □

This theorem can be extended to the case of a singular $\boldsymbol{\Sigma}$ (cf. Cambanis, Huang and Simons, 1981).

In general, a given variable $\mathbf{x} \sim EC_n(\boldsymbol{\mu}, \boldsymbol{\Sigma}, \phi)$ does not necessarily possess a density (cf. Section 2.3 on spherical distributions). We shall now consider two cases: (1) $\mathbf{x}$ has a density; (2) $\boldsymbol{\Sigma} > \mathbf{0}$ and $P(\mathbf{x} = \boldsymbol{\mu}) = 0$.

Clearly, a necessary condition that $\mathbf{x} \sim EC(\boldsymbol{\mu}, \boldsymbol{\Sigma}, \phi)$ possesses a density is that $\operatorname{rank}(\boldsymbol{\Sigma}) = n$. In this case, the stochastic representation becomes

$$\mathbf{x} \overset{d}{=} \boldsymbol{\mu} + \mathbf{A}'\mathbf{y}$$

where $\mathbf{A}$ is a nonsingular matrix with $\mathbf{A}'\mathbf{A} = \boldsymbol{\Sigma}$ and $\mathbf{y} \sim S_n(\phi)$ (cf. (2.10)).

The density of $\mathbf{y}$ is of the form $g(\mathbf{y}'\mathbf{y})$, where $g(\cdot)$ is the density generator. Since $\mathbf{x} = \boldsymbol{\mu} + \mathbf{A}'\mathbf{y}$, the density of $\mathbf{x}$ is of the form

$$|\boldsymbol{\Sigma}|^{-1/2} g((\mathbf{x} - \boldsymbol{\mu})'\boldsymbol{\Sigma}^{-1}(\mathbf{x} - \boldsymbol{\mu})). \tag{2.43}$$

In this case we shall sometimes use the notation $EC_n(\boldsymbol{\mu}, \boldsymbol{\Sigma}, g)$ instead of $EC_n(\boldsymbol{\mu}, \boldsymbol{\Sigma}, \phi)$.

If $\mathbf{x}$ does not possess a density, $P(r = 0) = 0$ (cf. (2.33)) and $\boldsymbol{\Sigma} > 0$, carrying out the same transformation $\mathbf{x} = \boldsymbol{\mu} + \mathbf{A}'\mathbf{y}$, we obtain that $P(\mathbf{y} = \mathbf{0}) = 0$ and $\mathbf{y}$ has all the marginal densities and so does $\mathbf{x}$. In this case, by (2.22) the marginal density of $\mathbf{x}_{(k)} = (X_1, \ldots, X_k) - \boldsymbol{\mu}_{(k)}$, $1 \leqslant k < n$, where $\boldsymbol{\mu}_{(k)} = (\mu_1, \ldots, \mu_k)'$, is given by

$$\begin{aligned}&\frac{\Gamma(n/2)|\boldsymbol{\Sigma}_k|^{1/2}}{\Gamma((n-k)/2)\pi^{k/2}} \int_{(\mathbf{x}'_{(k)}\boldsymbol{\Sigma}_k^{-1}\mathbf{x}_{(k)})^{1/2}}^{\infty} r^{-(n-2)} \\ &\times (r^2 - \mathbf{x}'_{(k)}\boldsymbol{\Sigma}_k^{-1}\mathbf{x}_{(k)})^{(n-k)/2-1}\,\mathrm{d}F(r)\end{aligned} \tag{2.44}$$

where $\boldsymbol{\Sigma}_k$ is the first principal minor of $\boldsymbol{\Sigma}$ of dimension k.

Any function $g(\cdot)$ satisfying (2.19) defines a density (2.43) of an elliptically symmetric distribution with a normalizing constant C_n, where

$$C_n = \frac{\Gamma(n/2)}{2\pi^{n/2}\int_0^\infty r^{n-1} g(r^2)\,\mathrm{d}r}. \tag{2.45}$$

Example 2.6
For r, $s > 0$, $2N + n > 2$, let

$$g(t) = t^{N-1}\exp(-rt^s). \tag{2.46}$$

Equation (2.45) implies that

$$C_n = s\pi^{-n/2} r^{(2N+n-2)/(2s)}\Gamma(n/2)/\Gamma\left[\frac{2N+n-2}{2s}\right]. \tag{2.47}$$

The multivariate normal distribution is the special case $n = 1$, $s = 1$, $r = 1/2$. The case $s = 1$ was introduced and studied by Kotz (1975). We shall discuss this type of elliptical distribution in detail in Section 3.2.

Nomakuchi and Sakata (1988a) derived the structure of the covariance matrix for an elliptical distribution when some special symmetric regions of R^n are assigned equal probabilities.

On the plane R^2, let A_i ($i = 1,2,3,4$), be the ith quadrant and L_j ($j = 1,\ldots,6$), be the ray originated from the origin at an angle of $(j-1)\pi/3$ from the positive direction of the X-axis. Let B_j($j = 1,\ldots,6$), be the region between L_j and L_{j+1}, where we use the convention $L_7 = L_1$.

Theorem 2.19
Let $\mathbf{x} \sim EC_2(\mathbf{0}, \boldsymbol{\Sigma}, \phi)$. The following two statements are true:

(i) $P(\mathbf{x} \in A_i) = 1/4$, $i = 1,2,3,4$, if and only if $\boldsymbol{\Sigma} = \mathrm{diag}\{a, b\}$, where $a > 0$, $b > 0$;
(ii) $P(\mathbf{x} \in B_i) = 1/6$, $i = 1,\ldots,6$, if and only if $\boldsymbol{\Sigma} = \sigma^2\mathbf{I}_2$, where $\sigma > 0$.

Theorem 2.20
Let σ denote a permutation $\{\sigma(1), \sigma(2), \ldots, \sigma(n)\}$ of $\{1, 2, \ldots, n\}$ and let

$$c(\sigma) = \{\mathbf{x} \in R^n \mid X_{\sigma(1)} \leqslant X_{\sigma(2)} \leqslant \cdots \leqslant X_{\sigma(n)}\}.$$

If $\mathbf{x} \sim EC_n(\mathbf{0}, \boldsymbol{\Sigma}, \phi)$, where $n \geqslant 3$, and if $P(\mathbf{x} \in c(\sigma)) = 1/n!$ for all permutations, $\boldsymbol{\Sigma}$ must be of the form

$$\boldsymbol{\Sigma} = \alpha I_n + (\mathbf{a}\mathbf{1}_n' + \mathbf{1}_n\mathbf{a}'), \quad \text{where} \quad \alpha > 0,\ \mathbf{a} \in R^n.$$

For the proofs of these theorems the readers is referred to Nomakuchi and Sakata (1988a). See also Nomakuchi and Sakata (1988b).

Some inequalities for multinormal distributions, such as Slepian's inequality, can be extended to elliptical distributions. See Das Gupta *et al.* (1972) and Gordon (1987) and the references therein.

2.6 Mixtures of normal distributions

Let $n(\mathbf{x};\mathbf{0},\mathbf{\Sigma})$ be the density of $\mathbf{y}\sim N_n(\mathbf{0},\mathbf{\Sigma})$ and $w(v)$ be a weighting function, i.e., $w(v)\geqslant 0$ is increasing and

$$\int_0^\infty \mathrm{d}w(v)=1.$$

The so-called **mixture** of normal distributions associated with the weight $w(\cdot)$ is defined by its p.d.f.

$$\begin{aligned} f(\mathbf{x}) &= \int_0^\infty n\left(\mathbf{x};\mathbf{0};\frac{1}{v^2}\mathbf{I}\right)\mathrm{d}w(v) \\ &= (2\pi)^{-n/2}\int_0^\infty \exp\left(-\tfrac{1}{2}\mathbf{x}'\mathbf{x}/v^2\right)\mathrm{d}w(v), \end{aligned} \tag{2.48}$$

and hence $\mathbf{x}$ has an elliptical distribution. Therefore, the class of mixtures of normal distributions is a subclass of that of elliptical distributions. It is easy to verify that an n-dimensional random vector $\mathbf{x}$ has a density $f(\mathbf{x})$ of the form (2.48) if and only if $\mathbf{x}$ can be decomposed as:

$$\mathbf{x}=w\mathbf{y}, \tag{2.49}$$

where $\mathbf{y}\sim N(\mathbf{0},\mathbf{\Sigma})$ and w are independent, $w\geqslant 0$ and has the c.d.f. $w(v)$.

The main result is that $\mathbf{x}\sim EC_n(\mathbf{0},\mathbf{\Sigma},\phi)$ is a mixture of normal distributions if and only if $\phi\in\Phi_\infty$, as defined in (2.4). For simplicity, we shall consider in this section only the case of $\mathbf{\Sigma}=\mathbf{I}$.

Theorem 2.21

A function $\phi\in\Phi_\infty$ if and only if

$$\phi(x)=\int_0^\infty e^{-xr^2}\,\mathrm{d}F_\infty(r), \tag{2.50}$$

where $F_\infty(\cdot)$ is a c.d.f. over $(0,\infty)$.

PROOF Let $\phi \in \Phi_\infty$. This means that $(x_1, \ldots, x_n)$ has a spherical distribution for $n \geqslant 1$. Thus the sequence (x_i) is exchangeable, and the well-known theorem of de Finetti (Feller, 1971, Ch. VII, and references therein) asserts that there exists a σ-field $\mathscr{F}$ of events conditional upon which the x_i are independent and have the same distribution function F, say. Write

$$\Phi(t) = \int_{-\infty}^{\infty} e^{itx}\, \mathrm{d}\mathscr{F}(x) = E(e^{itx_n} | \mathscr{F}), \tag{2.51}$$

so that Φ is a random, but $\mathscr{F}$-measurable, continuous function and

$$\Phi(-t) = \overline{\Phi(t)}, \quad |\Phi(t)| \leqslant 1, \quad \Phi(0) = 1. \tag{2.52}$$

The conditional independence means that, for all real $t_1, \ldots, t_n$,

$$E\left\{\exp\left(i \sum_{r=1}^{n} t_r x_r\right) \middle| \mathscr{F}\right\} = \prod_{r=1}^{n} \Phi(t_r), \tag{2.53}$$

so that

$$\phi(t_i^2 + \cdots + t_n^2) = E\left\{\exp\left(i \sum_{r=1}^{n} t_r x_r\right)\right\} = E\left\{\prod_{r=1}^{n} \Phi(t_r)\right\}. \tag{2.54}$$

The spherical symmetry implies that the left-hand side of (2.54), and thus also the right-hand side, depend only on $t_1^2 + \cdots + t_n^2$. For any real u and v, write $t = (u^2 + v^2)^{1/2}$, and use (2.52) and (2.54) to compute

$$\begin{aligned} &E\{|\Phi(t) - \Phi(u)\Phi(v)|^2\} \\ &\quad = E[\{\Phi(t) - \Phi(u)\Phi(v)\}\{\Phi(-t) - \Phi(-u)\Phi(-v)\}] \\ &\quad = E\{\Phi(t)\Phi(-t)\} - E\{\Phi(t)\Phi(-u)\Phi(-v)\} - E\{\Phi(u)\Phi(v)\Phi(-t)\} \\ &\qquad + E\{\Phi(u)\Phi(v)\Phi(-u)\Phi(-v)\}. \end{aligned}$$

The four terms in this last expression are all of the form of the right-hand side of (2.54), and since $t = (u^2 + v^2)^{1/2}$ we have

$$\begin{aligned} E\{\Phi(t)\Phi(-t)\} &= \phi(2t^2) \\ E\{\Phi(t)\Phi(-u)\Phi(-v)\} &= \phi(t^2 + u^2 + v^2) = \phi(2t^2) \\ E\{\Phi(u)\Phi(v)\Phi(-t)\} &= \phi(u^2 + v^2 + t^2) = \phi(2t^2) \\ E\{\Phi(u)\Phi(v)\Phi(-u)\Phi(-v)\} &= \phi(2u^2 + 2v^2) = \phi(2t^2). \end{aligned} \tag{2.55}$$

Hence all the expressions on the left-hand side of (2.55) are the same, and

$$E\{|\Phi(t) - \Phi(u)\Phi(v)|^2\} = 0$$

or

$$\Phi(t) = \Phi(u)\Phi(v)$$

with probability one. Thus Φ satisfies, with probability one, the functional equation

$$\Phi\{(u^2 + v^2)^{1/2}\} = \Phi(u)\Phi(v)$$

for all rational u and v, and hence by continuity for all real u and v. Hence (see e.g. Feller, 1971, Ch. III), $\Phi(t) = \exp(-at^2/2)$ for some complex a, and (2.52) implies that a is real and nonnegative. Since $a = -2\log\Phi(1)$, a is an $\mathscr{F}$-measurable random variable, and thus

$$E(z|a) = E\{E(z|\mathscr{F})|a\}$$

for any random variable z. Setting

$$z = \exp\left(i\sum_{j=1}^{n} t_j x_j\right)$$

and using (2.53), we obtain

$$E\left\{\exp\left(i\sum_{j=1}^{n} t_j x_j|a\right)\right\} = E\left(\prod_{j=1}^{n} e^{-at_j^2/2}|a\right)$$

$$= \prod_{j=1}^{n} e^{-at_j^2/2} = \exp\left(-a\sum_{j=1}^{n} t_j^2/2\right).$$

Hence, conditional on a, $x_j\,(j = 1,\ldots,n)$ are independently distributed as $N(0,a)$. Let $F(\cdot)$ be the c.d.f. of a, then

$$E\left\{\exp\left(i\sum_{j=1}^{n} t_j x_j\right)\right\} = \int_0^\infty \exp\left(-a\sum_{j=1}^{n} t_j^2/2\right)\mathrm{d}F(a),$$

which completes the proof by setting $F_\infty(\cdot) = F(2\sqrt{\cdot})$. □

The original proof of Theorem 2.21 is due to Schoenberg (1938). The proof presented here is adapted from Kingman (1972).

Corollary

$\mathbf{x} \sim S_n(\phi)$ with $\phi \in \Phi_\infty$ if and only if

$$\mathbf{x} \overset{d}{=} r\mathbf{y}, \tag{2.56}$$

where $\mathbf{y} \sim N_n(\mathbf{0}, \mathbf{I}_n)$ is independent of $r \geqslant 0$. This implies that the distribution of $\mathbf{x}$ is a scale mixture of normal distributions.

As the structure (2.56) indicates, the mixture of normal distributions has the advantage of easily transferring properties enjoyed by normal distributions. Thus numerous results have appeared for mixtures of normal distributions rather than for elliptical distributions. Many properties of mixtures of normal distributions become very apparent when we view mixtures of normal distributions as elliptical distributions.

2.7 Robust statistics and regression model

2.7.1 Robust statistics

Let $\mathbf{x} = (X_1, \dots, X_n)'$ be an exchangeable random vector which can be viewed as a sample from a population with the distribution of X_1. In the case when $\mathbf{x}$ is normally distributed several statistics useful for inferential purposes, such as t- and F-statistics, are available. The following theorem shows that these statistics, being invariant under scalar multiplication, are robust in the class of spherical distributions.

Theorem 2.22
The distribution of a statistic $t(\mathbf{x})$ remains unchanged as long as $\mathbf{x} \sim S_n^+(\phi)$ (cf. Section 2.1) provided that

$$t(a\mathbf{x}) \stackrel{d}{=} t(\mathbf{x}) \qquad \text{for each } a > 0. \tag{2.57}$$

In this case $t(\mathbf{x}) \stackrel{d}{=} t(\mathbf{y})$ where $\mathbf{y} \sim N_n(\mathbf{0}, \mathbf{I}_n)$.

PROOF By assumption we have $\mathbf{x} \stackrel{d}{=} r\mathbf{u}^{(n)}$, $r \sim F(\cdot)$ and is independent of $\mathbf{u}^{(n)}$. Thus

$$\begin{aligned} t(\mathbf{x}) \stackrel{d}{=} t(r\mathbf{u}^{(n)}) &= \int_0^\infty t(r\mathbf{u}^{(n)} | R = r)\,\mathrm{d}F(r) \\ &= \int_0^\infty t(r\mathbf{u}^{(n)})\,\mathrm{d}F(r) \stackrel{d}{=} \int_0^\infty t(\mathbf{u}^{(n)})\,\mathrm{d}F(r) \\ &= t(\mathbf{u}^{(n)}), \end{aligned}$$

which is independent of r (or ϕ). As $\mathbf{y} \sim S_n^+(\phi)$ with $\phi(u) = \exp(-\frac{1}{2}u)$ the second assertion follows. □

Example 2.7 (The classical t-statistic)
It is well known that the t-statistic for $\mathbf{x}=(x_1,\ldots,x_n)'$ is

$$t(\mathbf{x})=\sqrt{n}\bar{x}/S, \tag{2.58}$$

where

$$\bar{x}=\frac{1}{n}\sum_{i=1}^{n}x_i=\frac{1}{n}\mathbf{1}_n'\mathbf{x}$$

$$S^2=\frac{1}{n-1}\sum_{i=1}^{n}(x_i-\bar{x})^2=\frac{1}{n-1}\mathbf{x}'\mathbf{D}\mathbf{x}$$

$$\mathbf{D}=\mathbf{I}_n-\frac{1}{n}\mathbf{1}_n\mathbf{1}_n'.$$

Clearly $t(a\mathbf{x})=t(\mathbf{x})$ for each $a>0$. Thus the distribution of $t(\mathbf{x})$ remains unchanged in the class of spherical distributions with $P(\mathbf{x}=\mathbf{0})=0$, i.e., $t(\mathbf{x})\sim t_{n-1}$, t-distribution with $(n-1)$ degrees of freedom. Moreover, if

$$\mathbf{x}\overset{d}{=}r\mathbf{u}^{(n)}, \tag{2.59}$$

where $r>0$ (r and $\mathbf{u}^{(n)}$ are not necessarily independent), then the distribution of $t(\mathbf{x})$ is independent of r, i.e., $t(\mathbf{x})\sim t_{n-1}$. Dickey and Chen (1985) present a discussion of this model for Bayesian inference.

In this connection we pose the following *open problem.* Set

$$T=\{\mathbf{x}:\mathbf{x}\in R^n \text{ is exchangeable and } t(\mathbf{x})\sim t_{n-1}\}, \tag{2.60}$$

where $t(\mathbf{x})$ is defined by (2.58). We wish to know how large the class T is. Clearly the $\mathbf{x}$ defined in (2.59) belongs to T. Our conjecture is that the class T is wider and includes additional variables besides the $\mathbf{x}$ defined in (2.59).

Example 2.8 (The F-statistic)
Let $\mathbf{x}\sim S_n^+(\phi)$ and $\mathbf{A}$ and $\mathbf{B}$ be two orthogonal projection matrices such that $\mathbf{AB}=\mathbf{0}$, rank $\mathbf{A}=a$, rank $\mathbf{B}=b$ and $a+b<n$. The F-statistic

$$F=\frac{\mathbf{x}'\mathbf{A}\mathbf{x}/a}{\mathbf{x}'\mathbf{B}\mathbf{x}/b} \tag{2.61}$$

retains the same distribution $F(a,b)$ (F-distribution with a and b degrees of freedom) in the class of $S_n^+(\phi)$, $\phi\in\Phi_n$.

Example 2.9 (*Orthant probability and probability of a cone*)
Numerous authors have attempted to calculate the orthant probability

$$F_n = P(\mathbf{y} \geqslant \mathbf{0}) = P(Y_1 \geqslant 0, \ldots, Y_n \geqslant 0), \tag{2.62}$$

for $\mathbf{y} \sim N_n(\mathbf{0}, \mathbf{R})$, where $\mathbf{R} = (\rho_{ij})$ is the correlation matrix of $\mathbf{y}$. For example, the following particular cases are well known:

$$F_2 = \frac{1}{4} + \arcsin(\rho_{12}/2\pi)$$

$$F_3 = \frac{1}{8} + \frac{1}{4\pi}(\arcsin\rho_{12} + \arcsin\rho_{13} + \arcsin\rho_{23}).$$

Gupta (1963) provides a comprehensive review and a bibliography of probability integrals of multivariate normal variables. Steck (1962) presents a substantial review of results pertaining to orthant probabilities denoted by $F_n(\rho)$, in the equicorrelated case and computes $F_n(\rho)$ for $\rho = .5(.1).9$ and $n = 2(1)50$. See also Johnson and Kotz (1972) for additional results and references.

In the case when $\mathbf{x} \sim EC_n(\mathbf{0}, \mathbf{R}, \phi)$, $P(\mathbf{x} = \mathbf{0}) = 0$ and $\mathbf{R} > \mathbf{0}$ is the correlation matrix of $\mathbf{x}$, the orthant probabilities are straightforwardly calculated. Indeed, by an argument similar to one presented in the proof of Theorem 2.22 we obtain that

$$P(\mathbf{x} \geqslant \mathbf{0}) = P(\mathbf{y} \geqslant \mathbf{0})$$

for $\mathbf{y}$ presented in (2.62). This implies that the orthant probabilities are invariant in the class of $EC_n(\mathbf{0}, \mathbf{R}, \phi)$, $\phi \in \Phi_n$ with $P(\mathbf{x} = \mathbf{0}) = 0$.

Moreover, the same technique can be applied to calculate probabilities of a cone. Let S be a connective area on the unit sphere of R^n. A cone C associated with S is defined by

$$C = \{\mathbf{x} : \mathbf{x} \in R^n, \mathbf{x}/\|\mathbf{x}\| \in S\} \cup \{0\}.$$

Let $(r, \theta_1, \ldots, \theta_{n-1})$ be the spherical coordinates of $\mathbf{x}$ (cf. Theorem 2.11). A cone of a spherical structure is given by

$$C = \{(r, \theta_1, \ldots, \theta_{n-1}) : r \geqslant 0, 0 \leqslant \alpha_k \leqslant \theta_k \leqslant \alpha_k^* \leqslant \pi, k = 1, \ldots, n-1,$$
$$0 \leqslant \alpha_{n-1} \leqslant \theta_{n-1} \leqslant \alpha_{n-1}^* < 2\pi\}, \tag{2.63}$$

where α_k and α_k^* $(k = 1, \ldots, n-1)$ are given values. It is easy to verify that

$$P(\mathbf{x} \in C) = P(\mathbf{y} \in C) \tag{2.64}$$

for each cone, where $\mathbf{y} \sim N(\mathbf{0}, \mathbf{I}_n)$. Thus the probability of a cone is invariant in the above mentioned class. A further discussion on this property will be presented in Section 4.2.

2.7.2 Regression model

In this subsection we shall give a brief discussion of a regression model in the case of spherical distributions of errors. Consider the following regression model:

$$\begin{cases} \mathbf{y} = \mathbf{X}\boldsymbol{\beta} + \mathbf{e} & \mathbf{y}: n \times 1,\ \mathbf{X}: n \times p,\ \boldsymbol{\beta}: p \times 1 \\ \mathbf{e} \sim \sigma^2 S_n^+(\phi), & \sigma^2 \text{ unknown.} \end{cases} \tag{2.65}$$

We shall assume for simplicity that $\operatorname{rank}(\mathbf{X}) = p < n$. It is well known that the least-squares estimator $\hat{\boldsymbol{\beta}}$ of $\boldsymbol{\beta}$ is given by

$$\hat{\boldsymbol{\beta}} = (\mathbf{X}'\mathbf{X})^{-1}\mathbf{X}'\mathbf{y}. \tag{2.66}$$

Since $\mathbf{y} \sim EC_n(\mathbf{X}\boldsymbol{\beta}, \sigma^2\mathbf{I}_n, \phi)$, it is evident that $\hat{\boldsymbol{\beta}} \sim EC_p(\boldsymbol{\beta}, \sigma^2(\mathbf{X}'\mathbf{X})^{-1}, \phi)$. If $E(\mathbf{e}'\mathbf{e}) = \sigma^2$, $\hat{\boldsymbol{\beta}}$ is an unbiased estimator of $\boldsymbol{\beta}$. An unbiased estimator of σ^2 is

$$\hat{\sigma}^2 = \frac{1}{n-p}\mathbf{y}'(\mathbf{I} - \mathbf{X}(\mathbf{X}'\mathbf{X})^{-1}\mathbf{X}')\mathbf{y}. \tag{2.67}$$

For the testing of the hypotheses

$$H_0: \mathbf{H}\boldsymbol{\beta} = \mathbf{c}, \quad H_1: \mathbf{H}\boldsymbol{\beta} \neq \mathbf{c}, \tag{2.68}$$

where $\mathbf{H}: q \times p$, $q < p$, $\mathbf{c}: q \times 1$, Anderson and Fang (1982) obtained the following useful result.

Theorem 2.23
Consider the regression model (2.65) with $\mathbf{e} \sim \sigma^2 S_n(g)$ and g being a continuous decreasing function. Then the likelihood ratio criteria for testing hypotheses (2.68) is

$$\lambda = \frac{(\mathbf{H}\hat{\boldsymbol{\beta}} - \mathbf{c})'(\mathbf{H}(\mathbf{X}'\mathbf{X})^{-1}\mathbf{H}')^{-1}(\mathbf{H}\hat{\boldsymbol{\beta}} - \mathbf{c})}{\mathbf{y}'(\mathbf{I} - \mathbf{X}(\mathbf{X}'\mathbf{X})^{-1}\mathbf{X}')\mathbf{y}}. \tag{2.69}$$

Moreover, under the null hypothesis, $((n-p)/q)\lambda \sim F(q, n-p)$, which is independent of g.

Thus the normal theory for the parameter $\boldsymbol{\beta}$ remains valid in the case of the spherical model. For more details see Thomas (1970),

Fraser and Ng (1980), Nimmo-Smith (1979), Eaton (1985, 1986), Berk (1986), Kuritsyn (1986), Fang and Zhang (1989). In particular, Fraser and Ng (1980) have shown that the statistics for testing for $\boldsymbol{\beta}$ are not only distribution-robust, but also inference-robust. Berk (1986) argues that $\hat{\boldsymbol{\beta}}$ is a sufficient statistic iff $\mathbf{e} \sim N(\mathbf{0}, \sigma^2 \mathbf{I}_n)$ for some $\sigma^2 > 0$. Nimmo-Smith (1979) establishes the relationship between linear regression and sphericity. Eaton (1985, 1986) and Kuritsyn (1986) use different approaches of the least-squares procedure for estimation of $\boldsymbol{\beta}$. For optimality–robustness of the test (2.69), see Giri and Das (1988) and Giri (1988) and the references cited there.

Large-sample inference for the location parameters of a linear growth curve model (a model which is more general than the multivariate linear model), canonical correlations and linear discriminant functions, based on an i.i.d. elliptical sample, is considered by Khatri (1988). Haralick (1977) and Jajuga (1987) considered discrimination and classification problems for elliptical distributions.

2.8 Log-elliptical and additive logistic elliptical distributions

The lognormal distribution is useful for growth data and was studied thoroughly by Aitchison and Brown (1957) and Johnson and Kotz (1970). There is also a discussion of the multivariate lognormal distribution by M.E. Johnson (1987, Ch. 5). In the first part of this section we shall define the log-elliptical distribution which is an extension of the multivariate lognormal distribution and is due to Bentler, Fang and Wu (1988). In the second part we shall introduce a new family of distributions – additive logistic elliptical distributions, which are an extension of additive logistic normal distribution. The latter were defined and investigated by Aitchison (1986).

2.8.1 *Multivariate log-elliptical distribution*

Let R_+^n be the positive part of R^n and $\log \mathbf{w} = (\log w_1, \ldots, \log w_n)'$.

Definition 2.3
Let $\mathbf{w} \in R_+^n$ be a random vector. If $\log \mathbf{w} \sim EC_n(\boldsymbol{\mu}, \boldsymbol{\Sigma}, \phi)$ we say $\mathbf{w}$ has a **log-elliptical distribution** and denote it by $\mathbf{w} \sim LEC_n(\boldsymbol{\mu}, \boldsymbol{\Sigma}, \phi)$. When $\boldsymbol{\mu} = \mathbf{0}$ and $\boldsymbol{\Sigma} = \mathbf{I}$ we shall write $\mathbf{w} \sim LS_n(\phi)$ for simplicity.

Evidently log-elliptical distributions are a natural generalization of lognormal distribution.

Theorem 2.24
Assume that $\mathbf{w} \sim LEC_n(\boldsymbol{\mu}, \boldsymbol{\Sigma}, \phi)$ with $\boldsymbol{\Sigma} > \mathbf{0}$ and possesses a density. Then the density function is of the form

$$|\boldsymbol{\Sigma}|^{-1/2}\left(\prod_{i=1}^{n} w_i^{-1}\right)g((\log \mathbf{w} - \boldsymbol{\mu})'\boldsymbol{\Sigma}^{-1}(\log \mathbf{w} - \boldsymbol{\mu})). \qquad (2.70)$$

In this case we write $\mathbf{w} \sim LEC_n(\boldsymbol{\mu}, \boldsymbol{\Sigma}, g)$.

PROOF The existence of a density of $\mathbf{w}$ implies existence of a density of $\mathbf{v} = \log \mathbf{w}$. The latter is of the form

$$|\boldsymbol{\Sigma}|^{1/2} g((\mathbf{v} - \boldsymbol{\mu})'\boldsymbol{\Sigma}^{-1}(\mathbf{v} - \boldsymbol{\mu})).$$

The assertion follows by noting that the Jacobian $J(\mathbf{v} \to \mathbf{w}) = \prod_i^n w_i^{-1}$.

□

All marginal and conditional distributions of a log-elliptical distribution are also log-elliptical distributions. This follows directly from Theorems 2.16 and 2.18. For convenience of reference we state this fact in the following theorem.

Theorem 2.25
Let $\mathbf{w} \sim LEC_n(\boldsymbol{\mu}, \boldsymbol{\Sigma}, \phi)$ with $\boldsymbol{\Sigma} > \mathbf{0}$. Partition $\mathbf{w}, \boldsymbol{\mu}$ and $\boldsymbol{\Sigma}$ into

$$\mathbf{w} = \begin{pmatrix} \mathbf{w}^{(1)} \\ \mathbf{w}^{(2)} \end{pmatrix}, \quad \boldsymbol{\mu} = \begin{pmatrix} \boldsymbol{\mu}^{(1)} \\ \boldsymbol{\mu}^{(2)} \end{pmatrix}, \quad \boldsymbol{\Sigma} = \begin{pmatrix} \boldsymbol{\Sigma}_{11} & \boldsymbol{\Sigma}_{12} \\ \boldsymbol{\Sigma}_{21} & \boldsymbol{\Sigma}_{22} \end{pmatrix}, \qquad (2.71)$$

where $\mathbf{w}^{(1)}$: $m \times 1$, $\boldsymbol{\mu}^{(1)}$: $m \times 1$, $\boldsymbol{\Sigma}_{11}$: $m \times m$ and $0 < m < n$. Then

(i) $\mathbf{w}^{(1)} \sim LEC_m(\boldsymbol{\mu}^{(1)}, \boldsymbol{\Sigma}_{11}, \phi)$;
(ii) $(\mathbf{w}^{(1)}|\mathbf{w}^{(2)}) \sim LEC_m(\boldsymbol{\mu}_{1.2}, \boldsymbol{\Sigma}_{11.2}, \phi_{q(\exp \mathbf{w}^{(2)})})$, where $\boldsymbol{\mu}_{1.2}$, $\boldsymbol{\Sigma}_{11.2}$ and $\phi_{q(\mathbf{a})}$ are defined in (2.42).

The mixed moments of log-elliptical distribution have an elegant formula presented in the following theorem.

Theorem 2.26
Assume that $\mathbf{w} \sim LEC_n(\boldsymbol{\mu}, \boldsymbol{\Sigma}, \phi)$, $\mathbf{m} = (m_1, \ldots, m_n)' \in R^n$ and that the mixed moment of $E(\prod_1^n w_i^{m_i})$ exists. Then

$$E\left(\prod_{i=1}^{n} w_i^{m_i}\right) = \exp(\mathbf{m}'\boldsymbol{\mu})\phi(-\mathbf{m}'\boldsymbol{\Sigma}\mathbf{m}). \qquad (2.72)$$

PROOF From the existence of the mixed moments and the relationship between the moment generating function and the c.f. we have

$$E\left(\prod_1^n w_i^{m_i}\right) = E\left(\prod_1^n e^{m_i v_i}\right) \quad (\text{where } v_i = \log w_i)$$
$$= E(\exp(\mathbf{m}'\mathbf{v}))$$
$$= \exp(\mathbf{m}'\boldsymbol{\mu})\phi(-\mathbf{m}'\boldsymbol{\Sigma}\mathbf{m}).$$

The last step is due to $\mathbf{v} \sim EC_n(\boldsymbol{\mu}, \boldsymbol{\Sigma}, \phi)$. The proof is completed. □

For a vector $\mathbf{w}$ having a lognormal distribution with parameters $\boldsymbol{\mu}$ and $\boldsymbol{\Sigma}$, the mixed moment is given by (2.72).
In particular,

$$E(\mathbf{w}) = \exp(\boldsymbol{\mu} + \tfrac{1}{2}\operatorname{diag}\boldsymbol{\Sigma}) \qquad \left(\text{where diag } \mathbf{A} = \begin{bmatrix} a_{11} \\ \vdots \\ a_{nn} \end{bmatrix}\right),$$

$$\operatorname{Cov}(\mathbf{w}) = (v_{ij}),$$

with

$$v_{ij} = \exp(\mu_i + \mu_j + \tfrac{1}{2}\sigma_{ii} + \tfrac{1}{2}\sigma_{jj})(\exp(\sigma_{ij}) - 1).$$

It is well known that the mode of a univariate lognormal $LN(\mu, \sigma^2)$ is $\exp(\mu - \sigma^2)$ (see e.g. Johnson and Kotz, 1970, Vol. 1). We shall derive the mode of a $LEC_n(\boldsymbol{\mu}, \boldsymbol{\Sigma}, g)$. Let $f(\mathbf{w})$ be the density of $LEC_n(\boldsymbol{\mu}, \boldsymbol{\Sigma}, g)$ with g being differentiable. Then

$$f(\mathbf{w}) = |\boldsymbol{\Sigma}|^{-1/2}\left(\prod_1^n w_i^{-1}\right)g(\Delta), \tag{2.73}$$

where

$$\Delta = (\log \mathbf{w} - \boldsymbol{\mu})'\boldsymbol{\Sigma}^{-1}(\log \mathbf{w} - \boldsymbol{\mu}). \tag{2.74}$$

Simple calculations show that equation $\partial f/\partial \mathbf{w} = \mathbf{0}$ is equivalent to

$$2g'(\Delta)\left[\sum_{j=1}^N \sigma^{ij}(\log w_j - \mu_j)\right] = g(\Delta),$$

where $\boldsymbol{\Sigma}^{-1} = (\sigma^{ij})$ and $\boldsymbol{\mu} = (\mu_1, \ldots, \mu_n)'$, or

$$2g'(\Delta)[\boldsymbol{\Sigma}^{-1}\log \mathbf{w} - \boldsymbol{\Sigma}^{-1}\boldsymbol{\mu}] = g(\Delta)\mathbf{1}_n,$$

or

$$\log \mathbf{w} = \boldsymbol{\mu} + \frac{g(\Delta)}{2g'(\Delta)}\boldsymbol{\Sigma}\mathbf{1}. \tag{2.75}$$

Provided the value of $g(\Delta)/2g'(\Delta)$ is independent of $\mathbf{w}$, the mode of an $LEC_n(\boldsymbol{\mu}, \boldsymbol{\Sigma}, g)$ will be

$$\mathbf{w}_m = \exp\left(\boldsymbol{\mu} + \frac{g(\Delta)}{2g'(\Delta)}\boldsymbol{\Sigma}\mathbf{1}\right). \tag{2.76}$$

Some examples will illustrate the applicability of this result.

Example 2.10 (The multivariate lognormal distribution)
Here

$$g(x) = (2\pi)^{-n/2}\exp(-\tfrac{1}{2}x)$$

and

$$\frac{g(\Delta)}{2g'(\Delta)} = \frac{\exp(-\frac{1}{2}\Delta)}{-\exp(-\frac{1}{2}\Delta)} = -1$$

which is independent of $\mathbf{w}$. So the mode of the multivariate lognormal distribution is

$$\mathbf{w}_m = \exp\{\boldsymbol{\mu} - \boldsymbol{\Sigma}\mathbf{1}\}. \tag{2.77}$$

Example 2.11 (Pearson Type II distribution)
The symmetric multivariate Pearson Type II distribution will be defined and studied in some detail in Section 3.4. In this case we have

$$g(x) = C_n(1-x)^r, \quad 0 \leqslant x \leqslant 1, \quad r > -1$$

$$C_n = \frac{\Gamma(n/2 + r + 1)}{\Gamma(r+1)\pi^{n/2}}$$

and hence

$$\frac{g(\Delta)}{2g'(\Delta)} = \frac{C_n(1-\Delta)^r}{-2C_{nr}(1-\Delta)^{r-1}} = -\frac{1-\Delta}{2r}, \tag{2.78}$$

which does depend on $\mathbf{w}$. Thus in this moment we cannot get the mode using (2.75). Let $\mathbf{y} = \log\mathbf{w} - \boldsymbol{\mu}$. In view of (2.74), (2.75) and (2.76) we have

$$\Delta = \mathbf{y}'\boldsymbol{\Sigma}^{-1}\mathbf{y} = \left(\frac{g(\Delta)}{2g'(\Delta)}\right)^2 \mathbf{1}'\boldsymbol{\Sigma}\mathbf{1}$$

$$= \frac{\sigma}{4r^2}(1-\Delta)^2 = \frac{\sigma}{4r^2}(1 - 2\Delta + \Delta^2) \qquad (\text{denote } \sigma = \mathbf{1}'\boldsymbol{\Sigma}\mathbf{1})$$

or equivalently

$$\Delta^2 - 2\left(1 + \frac{2r^2}{\sigma}\right)\Delta + 1 = 0.$$

The solutions of this equation

$$\Delta = 1 + \frac{2r}{\sigma}(r \pm \sqrt{\sigma + r^2}).$$

Since by definition $0 \leqslant \Delta \leqslant 1$ is required, only the solution

$$\Delta_1 = 1 + \frac{2r}{\sigma}(r - \sqrt{\sigma + r^2})$$

should be considered, and the corresponding $g(\Delta_1)/(2g'(\Delta_1))$ is

$$\frac{-(1 - \Delta_1)}{2r} = \frac{1}{\sigma}(r - \sqrt{\sigma + r^2})$$

which is independent of $\mathbf{w}$. Thus the mode is

$$\mathbf{w}_m = \exp\left\{\boldsymbol{\mu} + \frac{1}{\sigma}(r - \sqrt{\sigma + r^2})\boldsymbol{\Sigma}\mathbf{1}\right\}. \tag{2.79}$$

The reader will find more examples in the problems at the end of this chapter.

2.8.2 *Additive logistic elliptical distributions*

Let

$$B_N = \{\mathbf{x}: \mathbf{x} \in R^N_+, \quad \|\mathbf{x}\| = x_1 + \cdots + x_N = 1\}.$$

Following Aitchison (1986) any vector $\mathbf{x} \in B_N$ will be called a **composition**. This type of data plays an important role in many disciplines, such as geochemical compositions of rocks, sediments at different depths, industrials products, etc.

Given a $\mathbf{x} \in B_N$, denote an $n = (N - 1)$-dimensional vector formed by the first n components of $\mathbf{x}$ by $\mathbf{x}_{-N}$. Let

$$\mathbf{y} = \log(\mathbf{x}_{-N}/x_N) = \begin{bmatrix} \log(x_1/x_N) \\ \vdots \\ \log(x_n/x_N) \end{bmatrix}. \tag{2.80}$$

Equation (2.80) yields a one-to-one transformation from $\mathbf{x} \in B_N$ to

$\mathbf{y} \in R^n$ and its inverse transformation is given by

$$\begin{cases} x_i = \exp(y_i) \Big/ \left\{1 + \sum_1^n \exp(y_i)\right\}, & i = 1, \ldots, n, \\ x_N = 1 - \sum_1^n x_i = 1 \Big/ \left\{1 + \sum_1^n \exp(y_i)\right\}. \end{cases} \tag{2.81}$$

Definition 2.4
A random vector $\mathbf{x} \in B_N$ is said to have an **additive logistic elliptical distribution** $ALE_n(\boldsymbol{\mu}, \boldsymbol{\Sigma}, \phi)$ iff $\mathbf{y} = \log(\mathbf{x}_{-N}/x_N) \sim EC_n(\boldsymbol{\mu}, \boldsymbol{\Sigma}, \phi)$.

Transformation (2.80) can be rewritten in the following useful form:

$$\mathbf{y} = \mathbf{F} \log \mathbf{x} \quad \text{with} \quad \mathbf{F} = (\mathbf{I}_n \colon -\mathbf{1}_n) \colon n \times N. \tag{2.82}$$

There is a close relationship between log-elliptical and additive logistic elliptical distributions. Given $\mathbf{w} \sim LEC_N(\boldsymbol{\nu}, \mathbf{V}, \phi)$ its **composition** $\mathbf{x}$ defined by

$$\mathbf{x} = C(\mathbf{w}) = \mathbf{w}/\|\mathbf{w}\| \quad (\text{where } \|\mathbf{w}\| = w_1 + \cdots + w_N) \tag{2.83}$$

has an additive logistic elliptical distribution. In view of

$$\begin{aligned} \mathbf{y} &= \log(\mathbf{x}_{-N}/x_N) = \log(\mathbf{w}_{-N}/w_N) \\ &= \mathbf{F} \log \mathbf{w} \sim EC_n(\boldsymbol{\mu}, \boldsymbol{\Sigma}, \phi), \end{aligned}$$

where

$$\boldsymbol{\mu} = \mathbf{F}\boldsymbol{\nu}, \quad \boldsymbol{\Sigma} = \mathbf{F}\mathbf{V}\mathbf{F}', \tag{2.84}$$

it follows that $\mathbf{x} \sim ALE_n(\boldsymbol{\mu}, \boldsymbol{\Sigma}, \phi)$. It is evident from definition of $\mathbf{F}$ that

$$\boldsymbol{\mu} = \begin{bmatrix} \nu_1 - \nu_N \\ \vdots \\ \nu_n - \nu_N \end{bmatrix} \quad \text{and} \quad \boldsymbol{\Sigma} = \mathbf{V}_{11} - \mathbf{1}_n \mathbf{v}' - \mathbf{v}\mathbf{1}_n' + \mathbf{V}_{NN}\mathbf{1}_n\mathbf{1}_n',$$

where

$$\mathbf{V} = \begin{pmatrix} \mathbf{V}_{11} & \mathbf{v} \\ \mathbf{v}' & \mathbf{v}_{NN} \end{pmatrix}, \quad \mathbf{V}_{11} \, n \times n, \quad \mathbf{v} \colon n \times 1.$$

The following theorem gives the density function for $ALE_n(\boldsymbol{\mu}, \boldsymbol{\Sigma}, \phi)$.

Theorem 2.27
Assume that $\mathbf{x} \sim ALE_n(\boldsymbol{\mu}, \boldsymbol{\Sigma}, \phi)$ has a density (i.e., $\mathbf{x}_{-N}$ has a density). Then the density of $\mathbf{x}_{-N}$ is of the form

$$|\boldsymbol{\Sigma}|^{-1/2} \prod_1^N x_i^{-1} g((\log(\mathbf{x}_{-N}/x_N) - \boldsymbol{\mu})' \boldsymbol{\Sigma}^{-1} (\log(\mathbf{x}_{-N}/x_N) - \boldsymbol{\mu})). \tag{2.85}$$

PROOF The existence of density of $\mathbf{x}$ implies that $\mathbf{y} = \log(\mathbf{x}_{-N}/x_N)$ has a density, i.e., $\mathbf{y} \sim EC_n(\boldsymbol{\mu}, \boldsymbol{\Sigma}, g)$ for some g. Hence the density of $\mathbf{y}$ is

$$|\boldsymbol{\Sigma}|^{-1/2} g((\mathbf{y}-\boldsymbol{\mu})'\boldsymbol{\Sigma}^{-1}(\mathbf{y}-\boldsymbol{\mu})).$$

As

$$y_i = \log \frac{x_i}{1 - x_1 - \cdots - x_n}, \quad x_i > 0, \quad \sum x_i < 1, \quad i = 1, \ldots, n,$$
$$x_N = 1 - x_1 - \cdots - x_n,$$

and

$$\frac{\partial y_i}{\partial x_i} = \frac{1}{x_i} + \frac{1}{x_N}, \quad \frac{\partial y_i}{\partial x_j} = \frac{1}{x_N}, \quad i \neq j,$$

the Jacobian $J(\mathbf{y} \to \mathbf{x}) = \prod_i^N x_i^{-1}$ and the assertion follows. □

We now consider the moments of $\mathbf{x} \sim ALE_n(\boldsymbol{\mu}, \boldsymbol{\Sigma}, \phi)$. In general it is not necessary that $\mathbf{x}$ possesses moments. Utilizing Definition 2.4, we can only calculate moments for $\{x_i/x_j\}$, but not for $\{x_i\}$.

It is easy to verify that

$$\begin{aligned}
E(\log(x_i/x_j)) &= \mu_i - \mu_j \\
\operatorname{Cov}(\log(x_i/x_k),\ \log(x_j/x_l)) &= \sigma_{ij} - \sigma_{il} - \sigma_{kj} + \sigma_{kl} \\
E(x_i/x_j) &= E(\exp(\log(x_i/x_j))) \\
&= E(\exp(y_i - y_j)) \\
&= \exp(\mu_i - \mu_j)\phi(2\sigma_{ij} - \sigma_{ii} - \sigma_{jj}) \\
\operatorname{Cov}(x_i/x_k, x_j/x_l) &= \operatorname{Cov}(e^{y_i - y_k}, e^{y_j - y_l}) \\
&= E(\exp(y_i - y_k + y_j - y_l)) \\
&\quad - E(\exp(y_i - y_k))E(\exp(y_j - y_l)) \\
&= \exp(\mu_i + \mu_j - \mu_k - \mu_l)[\phi(2\sigma_{ik} + 2\sigma_{il} + 2\sigma_{jk} + 2\sigma_{jl} \\
&\quad - 2\sigma_{ij} - 2\sigma_{kl} - 2\sigma_{ik} - \sigma_{ii} - \sigma_{jj} - \sigma_{kk} - \sigma_{ll}) \\
&\quad - \phi(2\sigma_{ik} - \sigma_{ii} - \sigma_{kk})\phi(2\sigma_{jl} - \sigma_{jj} - \sigma_{ll})]
\end{aligned}$$

for all i, j, k and $l = 1, \ldots, n$.

Since the distribution is defined by a log ratio, the marginal distribution should have some special meaning. Let $\mathbf{S}$ be a $M \times N$ $(M < N)$ selection matrix, i.e., a matrix with M elements 1, one in each row and at most one in each column, with the remaining elements 0. The vector $\mathbf{Sx}$ is a subvector of $\mathbf{x}$ and its composition, as defined by (2.83), $C(\mathbf{Sx}) = \mathbf{Sx}/\|\mathbf{Sx}\|$, is called a **subcomposition** of $\mathbf{x}$. Here the distribution of a subcomposition plays a role of marginal distribution.

Theorem 2.28

Assume that $\mathbf{x} \sim ALE_n(\boldsymbol{\mu}, \boldsymbol{\Sigma}, \phi)$ and $C(\mathbf{Sx})$ is a subcomposition of $\mathbf{x}$ with M components. Then $C(\mathbf{sx}) \sim ALE_M(\boldsymbol{\mu}_S, \boldsymbol{\Sigma}_S, \phi)$ with $m = M - 1$,

$$\boldsymbol{\mu}_S = \mathbf{F}_m \mathbf{S} \boldsymbol{\mu}^*, \quad \boldsymbol{\Sigma}_S = \mathbf{F}_m \mathbf{S} \boldsymbol{\Sigma}^* \mathbf{S}' \mathbf{F}_m'$$

$$\boldsymbol{\mu}^* = \begin{pmatrix} \boldsymbol{\mu} \\ 0 \end{pmatrix} : N \times 1, \quad \boldsymbol{\Sigma}^* = \begin{pmatrix} \boldsymbol{\Sigma} & \mathbf{0} \\ \mathbf{0}' & 0 \end{pmatrix} : N \times N \quad \text{and}$$

$$\mathbf{F}_m = (\mathbf{I}_m : -\mathbf{1}_m). \tag{2.86}$$

PROOF Using (2.82) we have

$$\begin{aligned} \mathbf{y}_S &= \mathbf{F}_m \log(\mathbf{Sx}) = \mathbf{F}_m \mathbf{S} \log \mathbf{x} \\ &= \mathbf{F}_m \mathbf{S} \log(\mathbf{x}/X_N) \\ &= \mathbf{F}_m \mathbf{S} \mathbf{y}^*, \end{aligned}$$

where

$$\mathbf{y} = \log(\mathbf{x}_{-N}/x_N) \sim EC_n(\boldsymbol{\mu}, \boldsymbol{\Sigma}, \phi), \quad \mathbf{y}^* = \begin{pmatrix} \mathbf{y} \\ 0 \end{pmatrix} : N \times 1.$$

It is easy to verify that $\mathbf{y}^* \sim EC_n(\boldsymbol{\mu}^*, \boldsymbol{\Sigma}^*, \phi)$. The assertion follows from Theorem 2.16. □

In general the order of variables in $\mathbf{x}$ is arbitrary. A natural problem is to investigate the effect of the order of these variables. Let $\mathbf{P}$ be an $N \times N$ permutation matrix. The problem reduces to determination of the distribution of $\mathbf{x}_P = \mathbf{Px}$.

Theorem 2.29

Assume that $\mathbf{x} \sim ALE_n(\boldsymbol{\mu}, \boldsymbol{\Sigma}, \phi)$ and $\mathbf{P}$ is an $N \times N$ permutation matrix. Then $\mathbf{x}_P = \mathbf{Px} \sim ALE_n(\boldsymbol{\mu}_P, \boldsymbol{\Sigma}_P, \phi)$ with

$$\boldsymbol{\mu}_P = \mathbf{FP}\boldsymbol{\mu}^*, \quad \boldsymbol{\Sigma}_P = \mathbf{FP}\boldsymbol{\Sigma}^*\mathbf{P}'\mathbf{F}', \tag{2.87}$$

where $\mathbf{F} = (\mathbf{I}_n : -\mathbf{1}_n)$, $\boldsymbol{\mu}^*$ and $\boldsymbol{\Sigma}^*$ are given in Theorem 2.28, or equivalently

$$\boldsymbol{\mu}_P = \mathbf{Q}_P \boldsymbol{\mu}, \quad \boldsymbol{\Sigma}_P = \mathbf{Q}_P \boldsymbol{\Sigma} \mathbf{Q}_P' \tag{2.88}$$

$$\mathbf{Q}_P = \mathbf{FPF}'(\mathbf{FF}')^{-1}. \tag{2.89}$$

PROOF Let $\mathbf{y}_P = \mathbf{F}\log \mathbf{x}_P$. If we can show that $\mathbf{y}_P \sim EC_n(\boldsymbol{\mu}_P, \boldsymbol{\Sigma}_P, \phi)$, the proof will be completed. We have

$$\begin{aligned}\mathbf{y}_P &= \mathbf{F}\log(\mathbf{P}\mathbf{x}) = \mathbf{F}\mathbf{P}\log \mathbf{x}\\ &= \mathbf{F}\mathbf{P}\log(\mathbf{x}/X_N) = \mathbf{F}\mathbf{P}\mathbf{y}^*,\end{aligned}$$

where $\mathbf{y}^*$ has the same meaning as in the preceding theorem. Thus $\mathbf{y}_P \sim ALE_n(\boldsymbol{\mu}_P, \boldsymbol{\Sigma}_P, \phi)$ and $\boldsymbol{\mu}_P$ and $\mathbf{Q}_P$ are given by (2.87).

Let $g(\mathbf{x}) = (\prod_1^N x_i)^{1/N}$ be the geometric mean of $\mathbf{x}$, $\mathbf{z} = \log(\mathbf{x}/g(\mathbf{x}))$ and $\mathbf{z}_P = \log(\mathbf{x}_P/g(\mathbf{x}_P))$. Then

$$\begin{aligned}\mathbf{y}_P &= \mathbf{F}\log \mathbf{x}_P = \mathbf{F}\log(\mathbf{x}_P/g(\mathbf{x}_P)) = F\log(\mathbf{x}_P/g(\mathbf{x}))\\ &= \mathbf{F}\mathbf{P}\log(\mathbf{x}/g(\mathbf{x})) = \mathbf{F}\mathbf{P}\mathbf{z}.\end{aligned}$$

Since $\mathbf{z} = \mathbf{G}\log \mathbf{x}$ with $\mathbf{G} = \mathbf{I}_N - (1/N)\mathbf{1}_N\mathbf{1}'_N$, by a direct calculation we obtain

$$\mathbf{G} = \mathbf{F}'(\mathbf{F}\mathbf{F}')^{-1}\mathbf{F}.$$

Thus

$$\begin{aligned}\mathbf{z} &= \mathbf{G}\log \mathbf{x} = \mathbf{F}'(\mathbf{F}\mathbf{F}')^{-1}\mathbf{F}\log \mathbf{x}\\ &= \mathbf{F}'(\mathbf{F}\mathbf{F}')^{-1}\mathbf{y}\end{aligned}$$

and

$$\mathbf{y}_P = \mathbf{F}\mathbf{P}\mathbf{F}'(\mathbf{F}\mathbf{F}')^{-1}\mathbf{y} = \mathbf{Q}_P\mathbf{y}$$

which implies (2.88) and (2.89). □

Finally we shall study conditional distributions for an additive logistic elliptical distribution. Let $\mathbf{x}$ be distributed according to $ALE_n(\boldsymbol{\mu}, \boldsymbol{\Sigma}, \phi)$ and $\mathbf{x}, \boldsymbol{\mu}, \boldsymbol{\Sigma}$ be partitioned into

$$\mathbf{x} = \begin{pmatrix}\mathbf{x}^{(1)}\\ \mathbf{x}^{(2)}\end{pmatrix}, \quad \boldsymbol{\mu} = \begin{pmatrix}\boldsymbol{\mu}^{(1)}\\ \boldsymbol{\mu}^{(2)}\end{pmatrix} \quad \text{and} \quad \boldsymbol{\Sigma} = \begin{pmatrix}\boldsymbol{\Sigma}_{11} & \boldsymbol{\Sigma}_{12}\\ \boldsymbol{\Sigma}_{21} & \boldsymbol{\Sigma}_{22}\end{pmatrix}, \tag{2.90}$$

where $\mathbf{x}^{(1)}: M \times 1$, $\mathbf{x}^{(2)}: (N-M) \times 1$, $\boldsymbol{\mu}^{(1)} = M \times 1$, $\boldsymbol{\mu}^{(2)}: (n-M) \times 1$, $\boldsymbol{\Sigma}_{11}: M \times M, m = M - 1$. Let $\mathbf{S}_1 = (\mathbf{I}_M : 0)$ and $\mathbf{S}_2 = (\mathbf{0} : \mathbf{I}_{N-M})$ be $M \times N$ and $(N-M) \times N$ selection matrices. Then $\mathbf{x}^{(i)} = \mathbf{S}_i\mathbf{x}$, $i = 1, 2$. Let $C(\mathbf{x}^{(1)})$ be the composition of $\mathbf{x}^{(1)}$, i.e.,

$$C(\mathbf{x}^{(1)}) = \mathbf{x}^{(1)}/\|\mathbf{x}^{(1)}\|.$$

We wish to obtain the conditional distribution of $C(\mathbf{x}^{(1)})$ given $\mathbf{x}^{(2)}$ which is equivalent to the conditional distribution of $\mathbf{v}_1 = \mathbf{F}_m \log C(\mathbf{x}^{(1)}) = \mathbf{F}_m \log \mathbf{x}^{(1)} = \mathbf{F}_m\mathbf{S}_1 \log \mathbf{x}$ given $\mathbf{v}_2 = \mathbf{F}_{n-M}\mathbf{S}_2 \log \mathbf{x}$, where

$\mathbf{F}_m = (\mathbf{I}_m \vdots -\mathbf{1}_m)$ and $\mathbf{F}_{n-M}$ are similarly defined. Thus

$$\begin{pmatrix}\mathbf{v}_1\\ \mathbf{v}_2\end{pmatrix} = \begin{pmatrix}\mathbf{F}_m\mathbf{S}_1\\ \mathbf{F}_{n-M}\mathbf{S}_2\end{pmatrix}\log\mathbf{x} = \begin{pmatrix}\mathbf{F}_m\mathbf{S}_1\\ \mathbf{F}_{n-M}\mathbf{S}_2\end{pmatrix}\log(\mathbf{x}/X_N)$$

$$= \begin{pmatrix}\mathbf{F}_m\mathbf{S}_1\\ \mathbf{F}_{n-M}\mathbf{S}_2\end{pmatrix}\mathbf{y}^* \sim EC_{n-1}(\mathbf{Q}\boldsymbol{\mu}^*, \mathbf{Q}\boldsymbol{\Sigma}^*\mathbf{Q}', \phi)$$

where $\boldsymbol{\mu}^*$ and $\boldsymbol{\Sigma}^*$ are as above, and

$$\mathbf{Q}\boldsymbol{\mu}^* = \begin{pmatrix}\mathbf{F}_m\boldsymbol{\mu}^{(2)}\\ \boldsymbol{\mu}^{(2)}\end{pmatrix}, \qquad \mathbf{Q}\boldsymbol{\Sigma}^*\mathbf{Q}' = \begin{pmatrix}\mathbf{F}_m\boldsymbol{\Sigma}_{11}\mathbf{F}_m' & \mathbf{F}_m\boldsymbol{\Sigma}_{12}\\ \boldsymbol{\Sigma}_{21}\mathbf{F}_m' & \boldsymbol{\Sigma}_{22}\end{pmatrix}.$$

From Theorem 2.18 the conditional distribution of $\mathbf{v}_1$ given $\mathbf{v}_2$ is $EC_m(\mathbf{F}_m\boldsymbol{\mu}_{1.2}, \mathbf{F}_m\boldsymbol{\Sigma}_{11.2}\mathbf{F}_m', \phi_{q(\mathbf{v}_2)})$, where $\boldsymbol{\mu}_{1.2}$, $\boldsymbol{\Sigma}_{11.2}$ and $\phi_{q(\mathbf{v}_2)}$ have the same meaning as in (2.42).

Summarizing the above discussion we arrive at the following theorem.

Theorem 2.30
Under the above assumptions and notation, the conditional distribution of $C(\mathbf{x}^{(1)})$ given $\mathbf{x}^{(2)}$ is

$$(C(\mathbf{x}^{(1)})|\mathbf{x}^{(2)}) \sim ALE_m(\mathbf{F}_m\boldsymbol{\mu}_{1.2}, \mathbf{F}_m\boldsymbol{\Sigma}_{11.2}\mathbf{F}_m', \phi_{q(\mathbf{v}^{(2)})}),$$

where $\mathbf{v}^{(2)} = \log\mathbf{x}^{(2)}/X_N$.

The material of this subsection is from Fang, Bentler and Chou (1988). The reader can find more interesting facts in the problems at the end of this chapter.

2.9 Complex elliptically symmetric distributions

So far we have treated elliptically symmetric distributions only in the case of real-valued random variables. However, complex multivariate distributions play an important role in time series, especially those in the frequency domain and in nuclear physics among other fields of application. A review of the literature on complex multivariate distributions is presented in Krishnaiah (1976). In fact, the theory of elliptical distributions in the complex case can be derived by similar techniques to those used in the real case.

Let $\mathbf{R} = (\mathbf{r}_{jk})$ be a $2n \times 2n$ matrix, and $\mathbf{C} = (c_{jk})$ be an $n \times n$ matrix with the entries $c_{jk} = a_{jk} + ib_{jk}$ and

$$\mathbf{r}_{jk} = \begin{pmatrix} a_{jk} & -b_{jk} \\ b_{jk} & a_{jk} \end{pmatrix}.$$

Then $\mathbf{R}$ is isomorphic to $\mathbf{C}$ and this is denoted as $\mathbf{R} \cong \mathbf{C}$.

We need the following known results (cf. Goodman, 1983):

1. If $\mathbf{R} \cong \mathbf{C}$, then $\alpha\mathbf{R} \cong \alpha\mathbf{C}$ for any $\alpha \in R$.
2. If $\mathbf{R} \cong \mathbf{C}$, then
   (a) $\mathbf{R}$ is symmetric $\Leftrightarrow$ $\mathbf{C}$ is Hermitian;
   (b) $\mathbf{R}$ is nonsingular $\Leftrightarrow$ $\mathbf{C}$ is nonsingular;
   (c) $\mathbf{R}$ is orthogonal $\Leftrightarrow$ $\mathbf{C}$ is unitary;
   (d) $|\mathbf{R}| = |\mathbf{C}|^2$;
   (e) if $\mathbf{C}$ is Hermitian, then $\operatorname{rank}(\mathbf{R}) = 2\operatorname{rank}(\mathbf{C})$;
   (f) $\mathbf{R}$ is symmetric positive definite $\Leftrightarrow$ $\mathbf{C}$ is Hermitian positive definite.

Let $\mathbf{z} = \mathbf{x} + i\mathbf{y}$ be an $n \times 1$ complex vector, where $\mathbf{x} = (x_1, \ldots, x_n)'$ and $\mathbf{y} = (y_1, \ldots, y_n)'$ are real, and let $\mathbf{z}_R = (x_1, y_1, \ldots, x_n, y_n)'$ denote the vector of the real and imaginary parts of $\mathbf{z}$.

Definition 2.5

A complex random vector $\mathbf{z}$ has a **complex elliptically symmetric** distribution if and only if $\mathbf{z}_R$ has an elliptically symmetric distribution, i.e.,

$$\mathbf{z} \sim EC_n(\boldsymbol{\mu}, \boldsymbol{\Sigma}, \phi) \Leftrightarrow \mathbf{z}_R \sim EC_{2n}(\boldsymbol{\mu}_R, \boldsymbol{\Sigma}_R, \phi),$$

where the vector $\boldsymbol{\mu}_R$ has the same structure as $\mathbf{z}_R$ and $2\boldsymbol{\Sigma}_R \cong \boldsymbol{\Sigma}$.

Since the c.f. of $\mathbf{z}_R$ is

$$\phi_{\mathbf{z}_R}(\mathbf{t}_R) = \exp\{i\mathbf{t}'_R\boldsymbol{\mu}_R\}\,\phi(\mathbf{t}'_R\boldsymbol{\Sigma}_R\mathbf{t}_R),$$

the c.f. of $\mathbf{z}$ is correspondingly

$$\phi_{\mathbf{z}}(\mathbf{t}) = \exp\{i\operatorname{Re}(\mathbf{t}^*\boldsymbol{\mu})\}\,\phi(\tfrac{1}{2}\mathbf{t}^*\mathbf{t}),$$

where the relation between $\mathbf{t}$ and $\mathbf{t}_R$ is the same as the one between $\mathbf{z}$ and $\mathbf{z}_R$; here $\mathbf{t}^*$ is the conjugate transpose of $\mathbf{t}$, and $\operatorname{Re}(\cdot)$ denotes the real part.

If $\mathbf{z}_R$ has a density

$$|\boldsymbol{\Sigma}_R|^{-1/2} f((\mathbf{z}_R - \boldsymbol{\mu}_R)'\boldsymbol{\Sigma}_R^{-1}(\mathbf{z}_R - \boldsymbol{\mu}_R)),$$

then the density of $\mathbf{z}$ is

$$2^n|\boldsymbol{\Sigma}|^{-1}f(2(\mathbf{z}-\boldsymbol{\mu})^*\boldsymbol{\Sigma}^{-1}(\mathbf{z}-\boldsymbol{\mu})).$$

The following important result deals with a stochastic representation of the random vector $\mathbf{z}$.

Theorem 2.31 $\mathbf{z} \sim EC_n(\boldsymbol{\mu}, \boldsymbol{\Sigma}, \phi)$ with $\operatorname{rank}(\boldsymbol{\Sigma}) = k$ if and only if

$$\mathbf{z} = \boldsymbol{\mu} + \frac{1}{\sqrt{2}} r\mathbf{A}^*\mathbf{u}^{(k)}$$

where $r \geqslant 0$ is independent of $\mathbf{u}^{(k)}$, $\mathbf{u}^{(k)}$ as defined above and $\boldsymbol{\Sigma} = \mathbf{A}^*\mathbf{A}$, $\mathbf{A}^*$ being the conjugate transpose of $\mathbf{A}$.

A proof can be found in Krishnaiah and Lin (1984) where the authors generalize the main results from the real to the complex case. Khatri (1987) also considers the complex elliptical distributions.

Problems

2.1 Let $f(\cdot)$ be a nonnegative function such that $f(\mathbf{x}'\mathbf{x})$ is a density function in R^n with the covariance matrix $\mathbf{I}_n$. Prove that $f(\cdot)$ satisfies the following normalizing conditions:

$$\int_0^\infty u^{n/2-1}f(u)\mathrm{d}u = \Gamma(n/2)\pi^{-n/2}$$

$$\int_0^\infty u^{n/2}f(u)\mathrm{d}u = \frac{2\Gamma(n/2+1)}{\pi^{n/2}}.$$

2.2 Let $f(\cdot)$ be a measurable function satisfying

$$\int_0^\infty y^{\alpha-1}|f(y)|\mathrm{d}y < \infty, \qquad \alpha > 0.$$

Define the operator

$$W^\alpha f(x) = \int_x^\infty (y-x)^{\alpha-1}f(y)\mathrm{d}y/\Gamma(\alpha), \qquad \alpha > 0, x > 0.$$

which is the so-called Weyl fractional integral operator (cf. Rooney, 1972). Prove that

(i) $W^{\alpha+\beta}f = W^\alpha W^\beta f$;

(ii) Formula (2.23) can be expressed as

$$g_m(u) = \pi^{(n-m)/2}W^{1/2(n-m)}g_n(u).$$

2.3 Assume that $\mathbf{x} = \begin{pmatrix} \mathbf{x}_{(1)} \\ \mathbf{x}_{(2)} \end{pmatrix} \sim S_n(g)$ with $\mathbf{x}_{(1)}: m \times 1$, $0 < m < n$. Prove that $(\mathbf{x}^{(1)} | \mathbf{x}^{(2)} = \mathbf{x}^0_{(2)}) \sim S_m(g_a)$, where $a = (\mathbf{x}^0_{(2)})'(\mathbf{x}^0_{(2)})$ and

$$g_a(u) = \frac{\Gamma(m/2)}{\pi^{m/2}} f(a+u) \Big/ \int_0^\infty u^{m/2-1} f(a+u)\,\mathrm{d}u,$$

or equivalently

$$g_a(u) = \frac{f(a+u)}{\pi^{m/2} W^{m/2} f(a+u)},$$

where W is the Weyl fractional integral operator defined in Problem 2.2.

2.4 Let $(X, Y)' = R\mathbf{u}^{(2)} \sim S_2(g)$ possessing the second moment. Denote $c(z) = E(X^2 | Y = z)$ and $e(z) = E(|X| \,|\, Y = z)$. Prove the following results:

(i) the distribution of (X, Y) can uniquely be determined by $c(z)$;

(ii) $g(z)c(z) = \int_z^\infty x g(x)\,\mathrm{d}x$;

(iii) $c(z) = \int_z^\infty (r^2 - z^2)^{1/2} \mathrm{d}G(r) / \int_z^\infty (r^2 - z^2)^{-1/2} \mathrm{d}G(r)$, where G is the c.d.f. of R;

(iv) $e(z)g(z) = (1 - G(z))/\pi$ (cf. Szablowski, 1986).

2.5 Let $\mathbf{x} \sim EC_n(\boldsymbol{\mu}, \boldsymbol{\Sigma}, g)$ be partitioned into

$$\mathbf{x} = \begin{pmatrix} \mathbf{x}_{(1)} \\ \mathbf{x}_{(2)} \end{pmatrix}, \quad \boldsymbol{\mu} = \begin{pmatrix} \boldsymbol{\mu}_{(1)} \\ \boldsymbol{\mu}_{(2)} \end{pmatrix}, \quad \boldsymbol{\Sigma} = \begin{pmatrix} \boldsymbol{\Sigma}_{11} & \boldsymbol{\Sigma}_{12} \\ \boldsymbol{\Sigma}_{21} & \boldsymbol{\Sigma}_{22} \end{pmatrix},$$

where $\mathbf{x}_{(1)}, \boldsymbol{\mu}_{(1)}: m \times 1$, $\boldsymbol{\Sigma}_{11}: m \times m$. Assume that $\mathbf{x}$ possesses the second moment. Prove that the conditional covariance matrix

$$\mathrm{Cov}(\mathbf{x}_{(1)} | \mathbf{x}_{(2)}) = \boldsymbol{\Sigma}_{11.2} C((\mathbf{x}_{(2)} - \boldsymbol{\mu}_{(2)})' \boldsymbol{\Sigma}_{22}^{-1} (\mathbf{x}_{(2)} - \boldsymbol{\mu}_{(2)})),$$

where $C: R_+ \to R_+$ satisfies the following equation:

$$C(X) g_{n-m}(x) = \int_x^\infty g_{n-m}(y)\,\mathrm{d}y,$$

where g_k is the k-dimensional density generator of $\mathbf{x}$ (cf. Szablowski, 1988a).

2.6 Prove Theorem 2.15 and provide an example to show that $\boldsymbol{\Sigma}, \phi, r$ and $\mathbf{A}$ are not unique.

2.7 Assume that $\mathbf{x} = (x_1, \ldots, x_N)' \sim ALE_n(\boldsymbol{\mu}, \boldsymbol{\Sigma}, \phi)$ and $\mathbf{y} = (y_1, \ldots, y_N)' \sim ALE_N(\boldsymbol{\nu}, \boldsymbol{\Omega}, \psi)$ are independent with $n = N - 1$.

Let $\mathbf{z} = (z_1, \ldots, z_N)'$ where

$$z_i = x_i y_i \bigg/ \sum_{j=1}^{N} x_j y_j, \quad i = 1, \ldots, N.$$

Show that $\mathbf{z} \sim ALE_n(\boldsymbol{\mu} + \boldsymbol{\nu}, \boldsymbol{\Sigma} + \boldsymbol{\Omega}, \phi\psi)$ (cf. Fang, Bentler and Chou, 1988, for this and related reproductive properties).

CHAPTER 3

Some subclasses of elliptical distributions

In this chapter, we shall discuss subclasses of elliptical distributions, such as multiuniform distributions, symmetric Kotz type distributions, symmetric multivariate Pearson Types VII and II distributions, and some others. These are the most useful subclasses from both practical and theoretical aspects. The well-known multinormal and multivariate t-distributions are classified as members of the above subclasses.

Table 3.1, which is essentially from Jensen (1985), presents a listing of these subclasses indicating the form of the density or c.f. Some calculations in this chapter were carried out by K.T. Fang, the first author of this book.

Table 3.1 *Some subclasses of n-dimensional spherical distributions*

Type	Density function $f(\mathbf{x})$ or c.f. $\psi(\mathbf{t})$
Kotz type	$f(\mathbf{x}) = c(\mathbf{x}'\mathbf{x})^{N-1}\exp(-r(\mathbf{x}'\mathbf{x})^s)$, $r, s > 0$, $2N + n > 2$
Multinormal	$f(\mathbf{x}) = c\exp(-\frac{1}{2}\mathbf{x}'\mathbf{x})$
Pearson Type VII	$f(\mathbf{x}) = c(1 + \mathbf{x}'\mathbf{x}/s)^{-N}$, $N > n/2$, $s > 0$
Multivariate t	$f(\mathbf{x}) = c(1 + \mathbf{x}'\mathbf{x}/s)^{-(n+m)/2}$, $m > 0$ an integer
Multivariate Cauchy	$f(\mathbf{x}) = c(1 + \mathbf{x}'\mathbf{x}/s)^{-(n+1)/2}$, $s > 0$
Pearson Type II	$f(\mathbf{x}) = c(1 - \mathbf{x}'\mathbf{x})^m$, $m > 0$
Logistic	$f(\mathbf{x}) = c\exp(-\mathbf{x}'\mathbf{x})/[1 + \exp(-\mathbf{x}'\mathbf{x})]^2$
Multivariate Bessel	$f(\mathbf{x}) = c(\lVert\mathbf{x}\rVert/\beta)^a K_a(\lVert\mathbf{x}\rVert/\beta)$, $a > -n/2$, $\beta > 0$, where $K_a(\cdot)$ denotes the modified Bessel function of the third kind
Scale mixture	$f(\mathbf{x}) = c\int_0^\infty t^{-n/2}\exp(-\mathbf{x}'\mathbf{x}/2t)\mathrm{d}G(t)$, $G(t)$ a c.d.f.
Stable laws	$\psi(\mathbf{t}) = \exp\{r(\mathbf{t}'\mathbf{t})^{\alpha/2}\}$, $0 < \alpha \leqslant 2$, $r < 0$
Multiuniform	$\psi(\mathbf{t}) = {}_0F_1(n/2; -\frac{1}{4}\lVert\mathbf{t}\rVert^2)$, ${}_0F_1(\cdot;\cdot)$ is a generalized hypergeometric function

3.1 Multiuniform distributions

Let $\mathbf{u}^{(n)}$ be uniformly distributed on the unit sphere surface in R^n which has been introduced in Section 2.1 and has frequently appeared in the subsequent development. We have seen that the 'uniform base' $\mathbf{u}^{(n)}$ plays an important role in the theory of spherical distributions and elliptically contoured distributions.

The multiuniform distributions are of substantial importance in statistical analysis of directional data (cf. Watson, 1984) as well as in projection pursuit methods (cf. Cheng, 1987). This section is devoted to a more detailed discussion of these multiuniform distributions on spheres.

3.1.1 The characteristic function

Let $\Omega_n(\|\mathbf{t}\|^2)$, $\mathbf{t}\in R^n$, be the c.f. of $\mathbf{u}^{(n)}$.

Theorem 3.1

The c.f. $\Omega_n(\|\mathbf{t}\|^2)$ can be expressed in the following equivalent forms:

$$\Omega_n(\|\mathbf{t}\|^2) = \frac{1}{B((n-1)/2, \frac{1}{2})} \int_0^{\pi} \exp(i\|\mathbf{t}\|\cos\theta)\sin^{n-2}\theta d\theta, \tag{3.1}$$

$$\Omega_n(\|\mathbf{t}\|^2) = \frac{\Gamma(n/2)}{\sqrt{\pi}\Gamma((n-1)/2)} \sum_{k=0}^{\infty} \frac{(-1)^k\|\mathbf{t}\|^{2k}}{(2k)!} \frac{\Gamma((n-1)/2)\Gamma((2k+1)/2)}{\Gamma((n+2k)/2)}, \tag{3.2}$$

$$\Omega_n(\|\mathbf{t}^2\|) = {}_0F_1\left(\frac{n}{2}; \frac{1}{4}\|\mathbf{t}\|^2\right), \tag{3.3}$$

where ${}_0F_1(\cdot;\cdot)$ is the so-called generalized hypergeometric function. Its general definition is

$${}_pF_q(a_1,\ldots,a_p; b_1,\ldots,b_q; z) = \sum_{k=0}^{\infty} \frac{a_1^{[k]}\cdots a_p^{[k]}}{b_1^{[k]}\cdots b_q^{[k]}} \frac{z_k}{k!} \tag{3.4}$$

where $a^{[k]}$ are the ascending factorials, $a^{[k]} = a(a+1)\cdots(a+k-1)$.

PROOF Since (cf. (2.9))

$$\mathbf{t}'\mathbf{u}^{(n)} \overset{d}{=} \|\mathbf{t}\| u_1, \qquad \mathbf{u}^{(n)} = (u_1,\ldots,u_n)'$$

and the density of u_1 is (cf. (1.26))

$$\frac{\Gamma(n/2)}{\Gamma((n-1)/2)\sqrt{\pi}}(1-u_1^2)^{(n-1)/2-1}, \quad u_1^2<1,$$

therefore we have

$$\begin{aligned}\Omega_n(\|\mathbf{t}\|^2) &= E(\exp i\mathbf{t}'\mathbf{u}^{(n)}) = E(\exp(i\|\mathbf{t}\|u_1))\\ &= \int_{-1}^{1}\exp(i\|\mathbf{t}\|u_1)\frac{\Gamma(n/2)}{\Gamma((n-1)/2)\sqrt{\pi}}(1-u_1^2)^{(n-1)/2-1}\,\mathrm{d}u_1\\ &= \frac{1}{B((n-1)/2,\frac{1}{2})}\int_0^{\pi}\exp(i\|\mathbf{t}\|\cos\theta)\sin^{n-2}\theta\mathrm{d}\theta,\\ &\quad (u_1=\cos\theta),\end{aligned}$$

and the representation (3.1) follows. Expression (3.2) follows from the integral

$$\int_0^{\pi}\cos^k\theta\cdot\sin^{n-2}\theta\mathrm{d}\theta=\begin{cases}B((n-1)/2,(k+1)/2), & \text{if } k \text{ is even}\\ 0 \qquad\qquad\qquad\qquad\quad, & \text{if } k \text{ is odd}\end{cases}$$

and relations between beta and gamma functions. In order to prove (3.3), we need only to verify the following lemma.

Lemma 3.1

$$\frac{\Gamma(n/2)}{\Gamma((n-1)/2)\sqrt{n}}\int_0^{\pi}\exp(z\cos\theta)\sin^{n-2}\theta\mathrm{d}\theta=\sum_{k=0}^{\infty}\frac{(\frac{1}{4}z^2)^k}{(\frac{1}{2})^{[k]}k!} \qquad (3.5)$$

$$={}_0F_1(\tfrac{1}{2}n;\tfrac{1}{4}z^2).$$

PROOF Let $I(n,z)$ denote the left-hand side of (3.5). Expand the exponential term in the integrand and integrate term by term. Noting that the terms corresponding to the odd powers of z are integrals of odd functions and hence vanish, we obtain

$$I(n,z)=\frac{\Gamma(n/2)}{\Gamma((n-1)/2)\sqrt{n}}\sum_{k=0}^{\infty}\frac{z^{2k}}{(2k)}\int_0^{\pi/2}\cos^{2k}\theta\sin^{n-2}\theta\,\mathrm{d}\theta.$$

Substituting $x=\sin^2\theta$, the integral on the right-hand side becomes

$$2\int_0^{\pi/2}\cos^{2k}\theta\cdot\sin^{n-2}\theta\,\mathrm{d}\theta=\int_0^1 x^{(n-3)/2}(1-x)^{k-1/2}\mathrm{d}x$$

$$=\frac{\Gamma((n-1)/2)(k+\frac{1}{2})}{\Gamma(k+(n/2))}.$$

Thus

$$I(n,z)=\sum_{k=0}^{\infty} z^{2k}\frac{\Gamma(\frac{1}{2}n)}{\Gamma(k+(n/2))}\frac{\Gamma(k+\frac{1}{2})}{(2k)!\sqrt{\pi}}.$$

Since

$$\frac{\Gamma(k+(n/2))}{\Gamma(n/2)}=(\tfrac{1}{2}n)^{[k]} \quad \text{and} \quad \frac{\Gamma(k+\frac{1}{2})}{(2k)!\sqrt{\pi}}=\frac{1}{4^k k!},$$

we arrive at the required result. □

Theorem 3.2
The function $\Omega_n(\|\mathbf{t}\|^2)$ with $\mathbf{t}\in R^{n+1}$ is not an $(n+1)$-dimensional characteristic function.

PROOF Suppose the contrary is true. Then there exists a distribution function F such that (cf. Theorem 2.2)

$$\Omega_n(u)=\int_0^{\infty}\Omega_{n+1}(ur^2)\mathrm{d}F(r), \quad u\geqslant 0$$

i.e., $\mathbf{u}^{(n)}\overset{d}{=}r\mathbf{u}_n$, where r is independent of $\mathbf{u}_n$, $r\geqslant 0$, and $\mathbf{u}_n$ is the subvector of $\mathbf{u}^{(n+1)}$ comprising the first n components. Since $\mathbf{u}_n$ has a p.d.f. (cf. (1.26)) and $P(r=0)=P(r\mathbf{u}_n=\mathbf{0})=P(\mathbf{u}^{(n)}=\mathbf{0})=0$, $\mathbf{u}^{(n)}$ possesses a density by Theorem 2.9. This contradiction completes the proof. □

This theorem, due to Fang and Chen (1984), shows that Φ_{n+1} is a proper subset of Φ_n for $n\geqslant 1$, where Φ_n is defined by (2.3).

3.1.2 Moments

Theorem 3.3
For any integers $m_1,\ldots,m_n$, with $m=\sum_{i=1}^n m_i$, the mixed moments of $\mathbf{u}^{(n)}$ can be expressed as:

$$E\left(\prod_{i=1}^n u_i^{m_i}\right)=\begin{cases}\dfrac{1}{(n/2)^{[l]}}\displaystyle\prod_{i=1}^n\frac{(2l_i)!}{4^{l_i}(l_i)!}, & \text{if } m_i=2l_i \text{ are even, } i=1,\ldots,n,\ m=2l;\\ 0\qquad\qquad, & \text{if at least one of the } m_i \text{ is odd.}\end{cases} \tag{3.6}$$

where as above $x^{[l]}=x(x+1)\cdots(x+l-1)$.

PROOF Let $\mathbf{x} \sim N_n(\mathbf{0}, \mathbf{I}_n)$. By (2.8) we have

$$\mathbf{x}/\|\mathbf{x}\| \stackrel{d}{=} \mathbf{u}^{(n)}$$

and $\|\mathbf{x}\|$ and $\mathbf{x}/\|\mathbf{x}\|$ are independent. Hence

$$E\left(\prod_{i=1}^{n} x_i^{m_i}\right) = E(\|\mathbf{x}\|^m)E\left(\sum_{i=1}^{n} u_i^{m_i}\right). \tag{3.7}$$

It is well known that the moments of the standard univariate normal distribution are given by:

$$E(x_1^s) = \begin{cases} 0 & , \quad \text{if } s \text{ is odd;} \\ \dfrac{s!}{2^{s/2}(s/2)!}, & \text{if } s \text{ is even.} \end{cases} \tag{3.8}$$

$$E(\|\mathbf{x}\|^s) = \frac{\Gamma((n+s)/2)2^{s/2}}{\Gamma(n/2)}.$$

Hence the assertion follows. □

Remark: Formula (3.6) is a special case of (2.18) with $r \equiv 1$. Indeed, for any spherical random variable $\mathbf{x} = (x_1, \ldots, x_n)'$, we have

$$\mathbf{x} \stackrel{d}{=} r\mathbf{u}^{(n)}$$

and

$$E\left(\prod_{i=1}^{n} x_i^{m_i}\right) = E(r^m)\cdot E\left(\prod_{i=1}^{n} u_i^{m_i}\right).$$

Now substitute (3.6) into the above formula and (2.18) is immediately obtained.

3.1.3 Marginal distribution

It is well known that all the marginal distributions of a multinormal distribution are also normal. However, the marginal distributions of $\mathbf{u}^{(n)}$ are not necessarily uniform. Indeed, the joint density of $(u_1, \ldots, u_k)$, $1 \leqslant k \leqslant n$, is given by

$$\frac{\Gamma(n/2)}{\Gamma((n-k)/2)\pi^{k/2}}\left(1 - \sum_{i=1}^{k} u_i^2\right)^{(n-k)/2-1}, \quad \text{if } \sum_{i=1}^{k} u_i^2 < 1, \tag{3.9}$$

which is a function of $\sum_{i=1}^{k} u_i^2$. Thus $\mathbf{u}^{(1)} = (u_1, \ldots, u_k)'$ has a spherical distribution whose stochastic decomposition is

$$\mathbf{u}^{(1)} = r_1 \mathbf{u}^{(k)},$$

where r_1 has the density (cf. (2.20))

$$(B((k/2),(n-k)/2))^{-1} 2r^{k-1}(1-r^2)^{(n-k)/2-1}, \quad 0<r<1.$$

Thus $r_1^2 \sim Be(k/2,(n-k)/2)$, a beta distribution with parameters $k/2$ and $(n-k)/2$. Similarly, $(u_1^2,\ldots,u_k^2) \sim D_{k+1}(\frac{1}{2},\ldots,\frac{1}{2};(n-k)/2)$.

3.1.4 *Conditional distributions*

Partition $\mathbf{u}^{(n)}$ into

$$\mathbf{u} = \begin{pmatrix} \mathbf{u}_{(1)} \\ \mathbf{u}_{(2)} \end{pmatrix}, \quad \text{where } \mathbf{u}_{(1)}: m \times 1.$$

Then, by Theorem 2.13, the conditional distribution $\mathbf{u}_{(1)}$ given $\mathbf{u}_{(2)} = \mathbf{a}$ is

$$(\mathbf{u}_{(1)} | \mathbf{u}_{(2)} = \mathbf{a}) \sim S_m(\Phi_{\|\mathbf{a}\|^2}) \tag{3.10}$$

with a stochastic representation

$$(\mathbf{u}_{(1)} | \mathbf{u}_{(2)} = \mathbf{a}) \overset{d}{=} r(\|\mathbf{a}\|^2)\mathbf{u}^{(m)}, \tag{3.11}$$

where for each $\|\mathbf{a}\|^2 \geqslant 0$, $r(\|\mathbf{a}\|^2)$ and $\mathbf{u}^{(m)}$ are independent, and

$$r(\|\mathbf{a}\|^2) \overset{d}{=} ((1-\|\mathbf{a}\|^2)^{1/2} | \mathbf{u}_{(2)} = \mathbf{a}) \overset{d}{=} (1-\|\mathbf{a}\|^2)^{1/2}, \tag{3.12}$$

which leads us to the following theorem.

Theorem 3.4
The conditional distribution of $\mathbf{u}_{(1)}$ given $\mathbf{u}_{(2)} = \mathbf{a}$ is a uniform distribution on the sphere with centre $\mathbf{0}$ and radius $(1-\|\mathbf{a}\|^2)^{1/2}$.

3.1.5 *Uniform distribution in the unit sphere*

Analogously to the uniform distribution on the surface of a unit sphere, the uniform distribution in a unit sphere in R^n is also a spherical distribution. Let $\mathbf{x}$ be distributed according to the uniform distribution in the unit sphere in R^n. Its density is given by (cf. Example 1.4)

$$P_{\mathbf{x}}(\mathbf{x}) = \begin{cases} n\Gamma(\frac{1}{2}n)/(2\pi^{n/2}), & \text{if } \sum_{i=1}^{n} x_i \leqslant 1 \\ 0 \quad , & \text{otherwise} \end{cases} \tag{3.13}$$

Clearly, $\mathbf{x}$ can be expressed as $\mathbf{x} \stackrel{d}{=} r\mathbf{u}^{(n)}$. The density of r is also given by (2.20), i.e.,

$$f(r) = \begin{cases} nr^{n-1} & \text{if } 0 \leqslant r \leqslant 1, \\ 0 & \text{otherwise.} \end{cases} \tag{3.14}$$

Thus, $r \sim Be(n, 1)$. From $f(r)$ and the density of $(u_1, \ldots, u_k)$, one can easily calculate the marginal density of $(x_1, \ldots, x_k)$, $1 \leqslant k < n$, which is

$$\frac{\Gamma((n+2)/2)}{\Gamma((n-k+2)/2)\pi^{k/2}} \left(1 - \sum_{i=1}^{k} x_i^2\right)^{(n-k)/2}, \quad \left(\sum_{i=1}^{k} x_i < 1\right).$$

By direct calculations the c.f. of $\mathbf{x}$ is $\phi(\mathbf{t}'\mathbf{t})$, where

$$\phi(u) = \frac{2}{B(\frac{1}{2}, (n+1)/2)} \int_0^\infty \cos(ux)(1 - x^2)^{(n-1)/2} \mathrm{d}x,$$

and for any integers $m_1, \ldots, m_n$ with $m = \sum_{i=1}^n m_i$, the mixed moments of $\mathbf{x}$ are

$$E\left(\prod_{i=1}^{n} x_i^{m_i}\right) = \frac{B(n+m, 1)}{B(n, 1)} \prod_{i=1}^{n} E(u_i^{m_i}).$$

The latter product is given by (3.6).

3.1.6 Discussion

The multiuniform distribution on a sphere plays an important role in the theory of directional distributions. Watson (1984) devotes a whole chapter in his book to this distribution where the reader can find many additional interesting results. Poincaré (1912) is to be credited for proving the fact that the first k coordinates of $\mathbf{u}^{(n)}$ are approximately independent normal. Diaconis and Freedman (1986) point out that the assertion still holds, even if $k = O(n)$ in the limit. This implies a finite form of de Finetti's theorem: an orthogonally invariant probability distribution in an n-dimensional space is almost a mixture of independent and identical distributions. This also explains why the spherical symmetry remains the most important property of independently and identically distributed normal distributions.

To meet the requirements of projection pursuit, Cheng (1987) considers the following problem. Let $\mathbf{x}_1, \ldots, \mathbf{x}_n$ be a sequence of p-dimensional i.i.d. random vectors, $\mathbf{x}_1 \sim F$, and $\mathbf{u} \stackrel{d}{=} \mathbf{u}^{(p)}$ be independent of $(\mathbf{x}_1, \ldots, \mathbf{x}_n)$. Let $\hat{F}_n^u(x)$ be the empirical distribution of a sequence $\mathbf{u}'\mathbf{x}_1, \ldots, \mathbf{u}'\mathbf{x}_n$. Under some regularity conditions Cheng obtains the limit distribution of $\hat{F}_n^{\mathbf{u}}(x)$ as both n and $p \to \infty$.

Jiang (1987) studies the limiting distribution of $\sum_{i=1}^k u_i z_i$, where $z_1, \ldots, z_k$ are points on the unit circle and $\mathbf{u} = (u_1, \ldots, u_k)' \sim D_k(\mathbf{b})$, a Dirichlet distribution. Under certain conditions the limiting distribution of $\sum_{i=1}^k u_i z_i$ is a uniform distribution on a unit circle.

3.2 Symmetric Kotz type distributions

3.2.1 *Definition*

Definition 3.1
If $\mathbf{x} \sim EC_n(\boldsymbol{\mu}, \boldsymbol{\Sigma}, g)$ and the density generator g is of the form

$$g(u) = C_n u^{N-1} \exp(-ru^s), \quad r, s > 0, 2N + n > 2, \tag{3.15}$$

where C_n is a normalizing constant, we say that $\mathbf{x}$ possesses a **symmetric Kotz type** distribution (cf. Kotz, 1975).

Thus, the density of $\mathbf{x}$ is given by

$$C_n |\boldsymbol{\Sigma}|^{-1/2} [(\mathbf{x} - \boldsymbol{\mu})'\boldsymbol{\Sigma}^{-1}(\mathbf{x} - \boldsymbol{\mu})]^{N-1} \exp\{-r[(\mathbf{x} - \boldsymbol{\mu})'\boldsymbol{\Sigma}^{-1}(\mathbf{x} - \boldsymbol{\mu})]^s\} \tag{3.16}$$

and using (2.45) we obtain

$$\begin{aligned} C_n &= \Gamma(n/2)\pi^{-n/2} \left\{ \int_0^\infty u^{n/2-1} u^{N-1} \exp(-ru^s) \, du \right\}^{-1} \\ &= \frac{s\Gamma(n/2)}{\pi^{n/2}\Gamma((2N+n-2)/2s)} r^{(2N+n-2)/2s}. \end{aligned} \tag{3.17}$$

When $N = 1$, $s = 1$ and $r = \frac{1}{2}$, the distribution reduces to a multinormal distribution. When $s = 1$, this is the original Kotz distribution introduced in Kotz (1975). This family of distributions was found to be useful in constructing models in which the usual normality assumption is not applicable (see, e.g. Koutras, 1986).

3.2.2 *Distribution of R^2*

Since the density of $U = R^2$, where R is the generating variate of $\mathbf{x}$, is given by

$$\frac{\pi^{n/2}}{\Gamma(n/2)} u^{n/2-1} g(u) \qquad \text{(cf. Section 2.2.3)}, \tag{3.18}$$

where $g(u)$ is defined by (3.15), the density of $U = R^2$ is thus given by

$$\left(sr^{(2N+n-2)/2s} \Big/ \Gamma\left(\frac{2N+n-2}{2s} \right) \right) u^{N+n/2-2} e^{-ru^s}, \quad u > 0. \quad (3.19)$$

For $s = 1$, the above density becomes

$$(r^{N+n/2-1}/\Gamma(N+n/2-1))u^{N+n/2-2}e^{-ru}, \quad u > 0, \quad (3.20)$$

which is a gamma distribution $\mathrm{Ga}(N+n/2-1, r)$. As the parameters of $\mathrm{Ga}(a, b)$ must be positive, the conditions of $2N - n > 2$ and $r > 0$ in (3.15) are essential.

3.2.3 *Moments*

Since the integral of (3.19) with respect to u over $[0, \infty)$ equals one, we have for any $t > 0$

$$\begin{aligned} E(R^{2t}) &= C_n \int_0^\infty u^t u^{N+n/2-2} e^{-ru^s} \, du \\ &= C_n / C_{n+2t} \\ &= r^{-t/s} \Gamma\left(\frac{2N+n+2t-2}{2s} \right) \Big/ \Gamma\left(\frac{2N+n-2}{2s} \right). \end{aligned}$$

In particular, we have

$$E(R^2) = r^{-1/s} \Gamma\left(\frac{2N+n}{2s} \right) \Big/ \Gamma\left(\frac{2N+n-2}{2s} \right)$$

and

$$E(R^2) = (N + n/2 - 1)/r$$

if $s = 1$. Now utilizing formula (2.18) it is easy to obtain any mixed moments of $\mathbf{x}$. In particular,

$$E(\mathbf{x}) = \boldsymbol{\mu} \quad \text{and} \quad D(\mathbf{x}) = r^{-1/s} \frac{\Gamma((2N+n)/2s)}{\Gamma((2N+n-2)/2s)} \boldsymbol{\Sigma}/\operatorname{rank} \boldsymbol{\Sigma}.$$

3.2.4 *Multivariate normal distributions*

For $N = s = 1$ and $r = \frac{1}{2}$ in (3.15), the function g becomes $g(u) = C_n \exp(-u/2)$ with $C_n = (2\pi)^{-n/2}$ which implies that $\mathbf{x} \sim N_n(\boldsymbol{\mu}, \boldsymbol{\Sigma})$. The appealing properties of multinormal distributions are discussed in numerous standard textbooks. Here we shall briefly discuss only a few features.

Conditional distributions

Assume that $\mathbf{x} \sim N_n(\boldsymbol{\mu}, \boldsymbol{\Sigma})$ with $\boldsymbol{\Sigma} > \mathbf{0}$ and $\mathbf{x}, \boldsymbol{\mu}$ and $\boldsymbol{\Sigma}$ are partitioned as in (2.38). Then the following result holds:

1. The conditional distribution of $\mathbf{x}^{(1)}$ given $\mathbf{x}^{(2)}$ is $N_m(\boldsymbol{\mu}_{1.2}, \boldsymbol{\Sigma}_{11.2})$ where $\boldsymbol{\mu}_{1.2}$ and $\boldsymbol{\Sigma}_{11.2}$ are as defined in (2.42).
2. $\mathbf{x}^{(1)} - \boldsymbol{\Sigma}_{12}\boldsymbol{\Sigma}_{22}^{-1}\boldsymbol{\Sigma}_{21}\mathbf{x}^{(2)}$ is independent of $\mathbf{x}^{(2)}$, and $\mathbf{x}^{(2)} - \boldsymbol{\Sigma}_{21}\boldsymbol{\Sigma}_{11}^{-1}\boldsymbol{\Sigma}_{12}\mathbf{x}^{(1)}$ is independent of $\mathbf{x}^{(1)}$.

Remark:
We can apply Theorem 2.18 directly to the normal case and prove result 1 above. Since $\mathbf{x} \sim EC_n(\boldsymbol{\mu}, \boldsymbol{\Sigma}, \boldsymbol{\Phi})$ with $\boldsymbol{\Phi}(u) = \exp(-u/2)$, we have that

$$(\mathbf{x}^{(1)} | \mathbf{x}^{(2)} = \mathbf{a}) \sim EC_n(\boldsymbol{\mu}_{1.2}, \boldsymbol{\Sigma}_{11.2}, \boldsymbol{\Phi}_{q(\mathbf{a})}),$$

where

$$\boldsymbol{\Phi}_{q(\mathbf{a})}(u) = \exp(-u/2)$$

is independent of $\mathbf{a}$. In Section 4.3 we shall point out that the fact that $\boldsymbol{\Phi}_{q(\mathbf{a})}$ does not depend on the value of $q(\mathbf{a})$ characterizes the normality of a distribution.

Moments

Withers (1985) obtained the following explicit expression for the moments of the multinormal distribution.

Theorem 3.5
For $\mathbf{x} \sim N_m(\boldsymbol{\mu}, \boldsymbol{\Sigma})$ and $i_1, \ldots, i_r$ belonging to the set $\{1, 2, \ldots, n\}$,

$$E(X_{i_1} \cdots X_{i_r}) = \sum_{l+2k=r} \sum^{m} \mu_{a_1} \cdots \mu_{a_l} \sigma_{b_1 b_2} \cdots \sigma_{b_{2k-1} b_{2k}},$$

where $\sum^m$ sums over all the $m = r!/(l! 2^k k!)$ permutations $(a_1, \ldots, a_l, b_1, b_2, \ldots, b_{2k})$ of $(i_1, \ldots, i_r)$ giving distinct terms allowing for the symmetry of $\boldsymbol{\Sigma}$, and $\sum_{l+2k=r}$ sums over all k and l with $l + 2k = r$. For illustrations see the examples below.

For the proof of this theorem and other explicit expressions in terms of multivariate Hermite polynomials, the reader is referred to Withers' paper.

Example 3.1
For $1 \leqslant i_1 < i_2 < i_3 \leqslant n$, we have $r = 3$; $l = 3$, $k = 0$ or $l = 1$, $k = 1$,

$$E(X_{i_1}X_{i_2}X_{i_3}) = \mu_{i_1}\mu_{i_2}\mu_{i_3} + \sum^{3} \mu_{a_1}\sigma_{b_1b_2}$$

$$= \mu_{i_1}\mu_{i_2}\mu_{i_3} + \mu_{i_1}\sigma_{i_2i_3} + \mu_{i_2}\sigma_{i_1i_3} + \mu_{i_3}\sigma_{i_1i_2},$$

where only one of the terms $\mu_{i_1}\sigma_{i_2i_3}$ and $\mu_{i_1}\sigma_{i_3i_2}$ appears in view of the symmetry of $\mathbf{\Sigma}$.

Example 3.2
For $1 \leqslant i_1 < i_2 \cdots < i_{2k} \leqslant n$, we have

$$E(X_{i_1} - \mu_{i_1})(X_{i_2} - \mu_{i_2})\cdots(X_{i_{2k}} - \mu_{i_{2k}}) = \sum^{m(k)} \sigma_{b_1b_2}\cdots\sigma_{b_{2k-1}b_{2k}},$$

where $\sum^{m(k)}$ is the sum over all $m(k) = (2k)!/(2^k k!) = 1.2.5\ldots(2k-1)$ permutations $(b_1, b_2, \ldots, b_{2k})$ of $(i_1, \ldots, i_{2k})$ giving distinct terms. This inductive formula was later generalized to elliptical distributions by Berkane and Bentler (1986) (see the discussion preceding Theorem 2.18). In particular,

$$E(X_i - \mu_i)(X_j - \mu_j)(X_k - \mu_k)(X_l - \mu_l) = \sigma_{ij}\sigma_{kl} + \sigma_{ik}\sigma_{jl} + \sigma_{il}\sigma_{jk},$$

which can also be computed directly by differentiating the c.f. (cf. Anderson, 1984, p. 49). Another formula for mixed moments of $N_n(\mathbf{0}, \mathbf{\Sigma})$ was developed by Guiard (1986). His development is based on an application of the moment generating function of a multivariate normal distribution. See also Holmquist (1988) for alternative formulas for moments and cumulants expressed in vector notation.

3.2.5 *The c.f. of Kotz type distributions*

For $s = 1$ in (3.15) we obtain a Kotz type distribution, the c.f. of which was recently derived by Iyengar and Tong (1988) as follows.

Theorem 3.6
The c.f. of the Kotz type distribution is

$$\exp(i\mathbf{t}'\boldsymbol{\mu})\psi_{n,N}(\mathbf{t}'\mathbf{\Sigma}\mathbf{t}; r),$$

where

$$\psi_{n,N}(u;r)=e^{-u/4r}\sum_{m=0}^{N-1}\binom{N-1}{m}\frac{\Gamma(n/2)}{\Gamma(n/2+m)}\left(-\frac{u}{4r}\right)^m. \quad (3.21)$$

PROOF Without loss of generality, we may assume $\boldsymbol{\Sigma}=\mathbf{I}$. Let

$$\lambda_{n,N}(\mathbf{t}'\mathbf{t};r)=\pi^{-n/2}\int e^{i\mathbf{t}'\mathbf{x}}(\mathbf{x}'\mathbf{x})^{N-1}\exp(-r(\mathbf{x}'\mathbf{x}))\mathrm{d}\mathbf{x}. \quad (3.22)$$

We shall prove by induction that

$$\lambda_{n,N}(u;r)=e^{-u/4r}\left(\frac{1}{r}\right)^{N+n/2-1}\sum_{m=0}^{N-1}\binom{N-1}{m}$$
$$\times\frac{\Gamma(N+n/2-1)}{\Gamma(n/2+m)}\left(-\frac{u}{4r}\right)^m. \quad (3.23)$$

Utilizing the c.f. of the multinormal, it is easy to show that (3.23) holds for $N=1$. Now assume that (3.23) is valid for N, we shall show that it holds for $N+1$. By differentiating (3.23) with respect to $(1/r)$, we have the recursive relation:

$$r^{-2}\frac{\mathrm{d}}{\mathrm{d}(1/r)}(\lambda_{n,N}(\mathbf{t}'\mathbf{t};r))$$
$$=r^{-2}\int e^{i\mathbf{t}'\mathbf{x}}(\mathbf{x}'\mathbf{x})^{N-1}\frac{\mathrm{d}}{\mathrm{d}(1/r)}(e^{-r(\mathbf{x}'\mathbf{x})})\,\mathrm{d}\mathbf{x}$$
$$=\int\exp(i\mathbf{t}'\mathbf{x})(\mathbf{x}'\mathbf{x})^N\exp(-r\mathbf{x}'\mathbf{x})\mathrm{d}\mathbf{x}=\lambda_{n,N+1}(\mathbf{t}'\mathbf{t};r)$$

i.e.,

$$\lambda_{n,N+1}(u;r)=r^{-2}\frac{\mathrm{d}}{\mathrm{d}(1/r)}\lambda_{n,N}(u;r).$$

Writing $k=-u/4r$ and $C_m=\binom{N-1}{m}\Gamma(N+n/2-1)/\Gamma(n/2+m)$ and carrying out the differentiation, we get

$$\lambda_{n,N+1}(u;r)=r^{-2}\frac{\mathrm{d}}{\mathrm{d}(1/r)}\left[e^k(1/r)^{N+n/2-1}\sum_{m=0}^{N-1}C_mk^m\right]$$
$$=e^kr^{-N-n/2}[k+(N+n/2-1)+m]\left[\sum_{m=0}^{N-1}C_mk^m\right]$$
$$=e^kr^{-N-n/2}\sum_{m=0}^{N-1}C_m(k^{m+1}+k^m(N+n/2+m-1))$$

$$= e^k r^{-N-n/2}\left[\sum_{v=1}^{N-1}\binom{N-1}{v-1}\frac{\Gamma(N+n/2-1)}{\Gamma(n/2+v-1)}k^v + k^N\right.$$

$$\left. + \sum_{m=0}^{N-1} C_m k^m (N+n/2+m-1)\right]$$

$$= e^k r^{-N-n/2}\left\{\sum_{m=0}^{N-1}\left[\binom{N-1}{m-1}\frac{\Gamma(N+n/2-1)}{\Gamma(n/2+m-1)}\right.\right.$$

$$\left.\left. + \binom{N-1}{m}\frac{\Gamma(N+n/2-1)}{\Gamma(n/2+m)}\left(N+\frac{n}{2}+m-1\right)\right]k^m + k^N\right\}$$

$$= e^k r^{-N-n/2}\sum_{m=0}^{N}\binom{N}{m}\frac{\Gamma(N+n/2)}{\Gamma(n/2+m)}k^m,$$

where in the last step we have made use of the identity

$$\binom{N-1}{m-1}\frac{\Gamma(N+n/2-1)}{\Gamma(n/2+m-1)}$$

$$+\binom{N-1}{m}\frac{\Gamma(N+n/2-1)}{\Gamma(n/2+m)}\times\left(N+\frac{n}{2}+m-1\right)$$

$$=\frac{N!}{m!(n-m)!}\left[\frac{m}{N}\frac{\Gamma(N+n/2-1)}{\Gamma(n/2+m-1)}\right.$$

$$\left.+\frac{N-m}{N}\frac{\Gamma(N+n/2-1)}{\Gamma(n/2+m)}\left(N+\frac{n}{2}+m-1\right)\right]$$

$$=\binom{N}{m}\Gamma(N+n/2)/\Gamma(n/2+m).$$

Thus formula (3.23) holds for $N+1$. The remaining calculations are straightforward and are left to the reader. □

3.3 Symmetric multivariate Pearson Type VII distributions

3.3.1 *Definition*

Definition 3.2

An $n \times 1$ random vector $\mathbf{x} \sim EC_n(\boldsymbol{\mu}, \boldsymbol{\Sigma}, g)$ is said to have a **symmetric multivariate Pearson Type VII** distribution if $\mathbf{x}$ has a density generator $g(\cdot)$, where

$$g(t) = C_n(1 + t/m)^{-N}, \qquad N > n/2, \quad m > 0 \tag{3.24}$$

and

$$C_n = (\pi m)^{-n/2}\Gamma(N)/\Gamma(N - n/2).$$

We shall denote this by $\mathbf{x} \sim MPVII_n(\boldsymbol{\mu}, \boldsymbol{\Sigma}, g)$.

Symmetric multivariate Pearson Type VII distributions include a number of important distributions such as the multivariate t-distribution (for $N = \frac{1}{2}(n + m)$) and the multivariate Cauchy distribution (for $m = 1$ and $N = \frac{1}{2}(n + 1)$).

Using (3.18), the density of $u = r^2$, where r is the generating variate of $\mathbf{x}$, is found to be

$$\frac{1}{B(n/2, N - n/2)} m^{-n/2} u^{n/2-1}(1 + u/m)^{-N}, \qquad u > 0, \qquad (3.25)$$

i.e., r^2/m has a beta type II distribution with parameters $n/2$ and $N - n/2$.

Remark:
A random variable B is said to have a beta type II distribution with parameters α and β if B has density

$$\frac{1}{B(\alpha, \beta)} b^{\alpha-1}(1 + b)^{-(\alpha+\beta)}, \qquad b > 0,$$

and we shall denote it by $B \sim \mathrm{BeII}(\alpha, \beta)$.

3.3.2 *Marginal densities*

Let $\mathbf{x} \stackrel{d}{=} \boldsymbol{\mu} + \boldsymbol{\Sigma}^{1/2}\mathbf{y}$ have a symmetric multivariate Pearson Type VII distribution $MPVII_n(\boldsymbol{\mu}, \boldsymbol{\Sigma}, g)$. Evidently $\mathbf{y} = MPVII_n(\mathbf{0}, \mathbf{I}, g)$, a spherical distribution. Partition $\mathbf{y}$ into

$$\mathbf{y} = \begin{pmatrix} \mathbf{y}^{(1)} \\ \mathbf{y}^{(2)} \end{pmatrix}, \quad \mathbf{y}^{(1)}: r \times 1. \qquad (3.26)$$

By formula (2.23), we have $\mathbf{y}^{(1)} \sim S_r(g_r)$ where the density generator $g_r(\cdot)$ is given by

$$g_r(u) = \frac{\pi^{(n-r)/2}}{\Gamma((n-r)/2)} \int_u^\infty (y - u)^{(n-r)/2-1} C_n(1 + y/m)^{-N} \mathrm{d}y$$

$$\text{(setting } x = y - u)$$

$$= \frac{\Gamma(N)\pi^{-r/2} m^{-n/2}}{\Gamma(N - n/2)\Gamma((n-r)/2)} \int_0^\infty x^{(n-r)/2-1}\left(1 + \frac{x + u}{m}\right)^{-N} \mathrm{d}x$$

$$= \frac{\Gamma(N)\pi^{-r/2}m^{-n/2}}{\Gamma(N-n/2)\Gamma((n-r)/2)}\left(1+\frac{u}{m}\right)^{-N}$$

$$\times \int_0^\infty x^{(n-r)/2-1}\left(1+\frac{x}{m+u}\right)^{-N} dx$$

$$\left(\text{setting } t = \frac{x}{m+u}\right)$$

$$= \frac{\Gamma(N-(n-r)/2)}{(\pi m)^{r/2}\Gamma(N-n/2)}\left(1+\frac{u}{m}\right)^{-N+(n-r)/2}.$$

If we use the notation $g_{N,m}(t)$ instead of $g(t)$ in (3.24), we would then have $g_r(t) = g_{N-(n-r)/2,m}(t)$, the same type of function. Summarizing the above result, we have the following theorem.

Theorem 3.7
Assume that $\mathbf{x} \sim MPVII_n(\boldsymbol{\mu}, \boldsymbol{\Sigma}, g_{N,m})$. Partition $\mathbf{x}, \boldsymbol{\mu}, \boldsymbol{\Sigma}$ in a similar manner into:

$$\mathbf{x} = \begin{pmatrix}\mathbf{x}^{(1)}\\ \mathbf{x}^{(2)}\end{pmatrix}, \quad \boldsymbol{\mu} = \begin{pmatrix}\boldsymbol{\mu}^{(1)}\\ \boldsymbol{\mu}^{(2)}\end{pmatrix}, \quad \boldsymbol{\Sigma} = \begin{pmatrix}\boldsymbol{\Sigma}_{11} & \boldsymbol{\Sigma}_{12}\\ \boldsymbol{\Sigma}_{21} & \boldsymbol{\Sigma}_{22}\end{pmatrix} \tag{3.27}$$

where $\mathbf{x}^{(1)}$, $\boldsymbol{\mu}^{(1)}$: $r \times 1$ and $\boldsymbol{\Sigma}_{11}$: $r \times r$. Then we have

$$\mathbf{x}^{(1)} \sim MPVII_r(\boldsymbol{\mu}^{(1)}, \boldsymbol{\Sigma}_{11}, g_{N-(n-r)/2,m}).$$

3.3.3 *Conditional distributions*

From Theorem 3.7 the density of $\mathbf{y}^{(2)}$ in (3.26) is

$$\frac{\Gamma(N-r/2)}{(\pi m)^{(n-r)/2}\Gamma(N-n/2)}\left(1+\frac{1}{m}\mathbf{y}^{(2)\prime}\mathbf{y}^{(2)}\right)^{-N+r/2}.$$

Let $u = \mathbf{y}^{(1)\prime}\mathbf{y}^{(1)}$ and $v = \mathbf{y}^{(2)\prime}\mathbf{y}^{(2)}$. The conditional density of $\mathbf{y}^{(1)}$ given $\mathbf{y}^{(2)}$ is

$$f(\mathbf{y}^{(1)}|\mathbf{y}^{(2)}) = \frac{\dfrac{\Gamma(N)}{(\pi m)^{n/2}\Gamma(N-n/2)}\left(1+\dfrac{1}{m}\mathbf{y}^{(1)\prime}\mathbf{y}^{(1)}+\mathbf{y}^{(2)\prime}\mathbf{y}^{(2)}\right)^{-N}}{\dfrac{\Gamma(N-r/2)}{(\pi m)^{(n-r)/2}\Gamma(N-n/2)}\left(1+\dfrac{1}{m}\mathbf{y}^{(2)\prime}\mathbf{y}^{(2)}\right)^{-N+r/2}}$$

$$= \frac{\Gamma(N)}{(\pi m)^{r/2}\Gamma(N-r/2)}\left(1+\frac{1}{m}v\right)^{-r/2}\left(1+\frac{u/m}{1+v/m}\right)^{-N}.$$

Let R_v be the generating variate of $(\mathbf{y}^{(1)}|\mathbf{y}^{(2)})$. Then the density of

$w = R_v^2/(m+v)$ is

$$\frac{\Gamma(N)}{\Gamma(r/2)\Gamma(N-r/2)} w^{r/2-1}(1+w)^{-N}.$$

We have thus established the following theorem.

Theorem 3.8
Under the same assumption as in Theorem 3.7, the conditional distribution of $\mathbf{x}^{(1)}$ given $\mathbf{x}^{(2)}$ has the following stochastic representation:

$$(\mathbf{x}^{(1)}|\mathbf{x}^{(2)}) \stackrel{d}{=} \boldsymbol{\mu}_{1.2} + R_v \boldsymbol{\Sigma}_{11.2}^{1/2}\mathbf{u}^{(r)},$$

where the vector $\boldsymbol{\mu}_{1.2}$ and the matrix $\boldsymbol{\Sigma}_{11.2}$ have the same meaning as in (2.42), $v = \mathbf{x}^{(2)\prime}\mathbf{x}^{(2)}$, the generating variate R_v of $\mathbf{x}^{(1)}|\mathbf{x}^{(2)}$ is independent of $\mathbf{u}^{(r)}$ for each v and $R_v^2/(m+v) \sim \text{BeII}(r/2, N-r/2)$.

3.3.4 *Moments*

Let $\mathbf{y} \sim MPVII_n(\mathbf{0}, \mathbf{I}_n, g_{N,m})$. Then

$$E(y_1^{s_1} \ldots y_n^{s_n}) = 0, \qquad \text{if some } s_i \text{ is odd,}$$

and

$$\begin{aligned} E\left(\prod_{j=1}^n y_j^{2s_j}\right) &= C_n \int_{R^n} \left(1 + \sum_{i=1}^n y_i^2/m\right)^{-N} \prod_{j=1}^n y_j^{2s_j} \mathrm{d}y_j \\ &= C_n \int_{R_+^n} \left(1 + \sum_1^n y_j/m\right)^{-N} \prod_{j=1}^n y_j^{s_j+1/2-1} \mathrm{d}y_j \\ &= C_n \left(\prod_{i=1}^n \Gamma(s_i + \tfrac{1}{2})/\Gamma\left(s + \frac{n}{2}\right)\right) \int_0^\infty z^{s+n/2-1} \\ &\quad \times (1 + z/m)^{-N} \mathrm{d}z \qquad (s = \Sigma s_j) \\ &= \frac{m^s \prod_1^n \Gamma(s_i + \frac{1}{2})\Gamma(N - s - n/2)}{\pi^{n/2}\Gamma(N - \frac{1}{2})}. \end{aligned} \tag{3.28}$$

The last step is obtained by using the definition of the beta type II distribution. In particular, we have

$$E(\mathbf{y}) = \mathbf{0},$$

$$\text{Cov}(\mathbf{y}) = \frac{m}{2N-n-2}\mathbf{I}_n,$$

and for the general case

$$E(\mathbf{x}) = \boldsymbol{\mu}, \qquad D(\mathbf{x}) = \frac{m}{2N - n - 2}\boldsymbol{\Sigma} \tag{3.29}$$

provided $\mathbf{x} \sim MPVII_n(\boldsymbol{\mu}, \boldsymbol{\Sigma}, g_{N,m})$ and $\boldsymbol{\Sigma} > \mathbf{0}$.

3.3.5 *Characteristic function*

Theorem 3.9
Let $\mathbf{x} \sim MPVII_n(\boldsymbol{\mu}, \boldsymbol{\Sigma}, g_{N,m})$. Then $\mathbf{x} \sim EC_n(\boldsymbol{\mu}, \boldsymbol{\Sigma}, \phi)$ with

$$\phi(u^2) = \frac{2\Gamma(N - (n-1)/2)}{\pi^{1/2}\Gamma(N - n/2)} \int_0^\infty \cos(m^{1/2}tu)(1 + t^2)^{-N + (n-1)/2}\,dt. \tag{3.30}$$

PROOF It is well known that if $\mathbf{x} \sim EC_n(\boldsymbol{\mu}, \boldsymbol{\Sigma}, \phi)$ and $\mathbf{x} \overset{d}{=} \boldsymbol{\mu} + \boldsymbol{\Sigma}^{1/2}\mathbf{y}$ with $\mathbf{y} \sim S_n(\phi)$, then $\phi(t^2) = E(e^{itY_1})$, where Y_1 is the first component of $\mathbf{y}$. For $\mathbf{x} \sim MPVII_n(\boldsymbol{\mu}, \boldsymbol{\Sigma}, g_{N,m})$, the density of Y_1 can be obtained directly as in Section 3.3.2. It is

$$\frac{\Gamma(N - (n-1)/2)}{(\pi m)^{1/2}\Gamma(N - n/2)}(1 + y_1^2/m)^{-N + (n-1)/2}.$$

Thus

$$\begin{aligned}\phi(t^2) &= \frac{\Gamma(N - (n-1)/2)}{(\pi m)^{1/2}\Gamma(N - n/2)} \int_{-\infty}^{\infty} e^{ity}(1 + y^2/m)^{-N + (n-1)/2}\,dy \\ &= \frac{2\Gamma(N - (n-1)/2)}{\pi^{1/2}\Gamma(N - n/2)} \int_0^\infty \cos(tm^{1/2}z)(1 + z^2)^{-N + (n-1)/2}\,dz,\end{aligned}$$

which completes the proof. □

3.3.6 *The multivariate t-distribution*

When $N = \frac{1}{2}(n + m)$ and m is an integer in (3.24), the corresponding distribution is called a **multivariate *t*-distribution** and is denoted by $\mathbf{x} \sim Mt_n(m, \boldsymbol{\mu}, \boldsymbol{\Sigma})$. In Example 2.5 we defined $Mt_n(m, \mathbf{0}, \mathbf{I}_n)$ by the stochastic representation. In fact, we can define $Mt_n(m, \boldsymbol{\mu}, \boldsymbol{\Sigma})$ in this manner. Let $\mathbf{z} \sim N_n(\mathbf{0}, \boldsymbol{\Sigma})$ and $S \sim \chi_m$ be independent. Then $\mathbf{x} \overset{d}{=} \boldsymbol{\mu} + m^{1/2}\mathbf{z}/s$ is said to have a multivariate t-distribution and is denoted as

$\mathbf{x} \sim Mt_n(m, \boldsymbol{\mu}, \boldsymbol{\Sigma})$. The equivalence between the two definitions is easy to establish. For example, by using the stochastic representation (for simplicity, assume $\boldsymbol{\Sigma} > \mathbf{0}$) we have

$$\mathbf{x} \overset{d}{=} \boldsymbol{\mu} + R\boldsymbol{\Sigma}^{1/2}\mathbf{u}^{(n)} \quad \text{for the first definition,}$$

$$\mathbf{x} \overset{d}{=} \boldsymbol{\mu} + R^*\boldsymbol{\Sigma}^{1/2}\mathbf{u}^{(n)} \quad \text{for the second definition.}$$

The proof that $R \overset{d}{=} R^*$ is straightforward and is left to the reader.

Applying the general theory of the preceding subsections to the multivariate t-distribution, we obtain the following useful results:

1. The density of $\mathbf{x} \sim Mt_n(m, \boldsymbol{\mu}, \boldsymbol{\Sigma})$ is

$$\frac{\Gamma((n+m)/2)}{(\pi m)^{n/2}\Gamma(m/2)}|\boldsymbol{\Sigma}|^{-1/2}(1 + m^{-1}(\mathbf{x}-\boldsymbol{\mu})'\boldsymbol{\Sigma}^{-1}(\mathbf{x}-\boldsymbol{\mu}))^{-(n+m)/2}.$$

A slightly different definition of the multivariate t-distribution is used by Johnson (1987). His version has the density function given by

$$\frac{\Gamma(m)}{\Gamma((m-n)/2)\pi^{n/2}}|\boldsymbol{\Sigma}|^{-1/2}(1 + (\mathbf{x}-\boldsymbol{\mu})'\boldsymbol{\Sigma}^{-1}(\mathbf{x}-\boldsymbol{\mu}))^{-m}.$$

2. If $\mathbf{x} \sim Mt_n(m, \boldsymbol{\mu}, \boldsymbol{\Sigma})$, and $\mathbf{y} = \mathbf{B}'\mathbf{x} + \boldsymbol{\nu}$, where $\mathbf{B}: k \times n$, and $\boldsymbol{\nu}: k \times 1$, then

$$\mathbf{y} \sim Mt_k(m, \mathbf{B}'\boldsymbol{\mu} + \boldsymbol{\nu}, \mathbf{B}'\boldsymbol{\Sigma}\mathbf{B}).$$

In particular, all the marginal distributions of a multivariate t- distribution are still multivariate t-distributions. If $\mathbf{x} \sim Mt_n(m, \boldsymbol{\mu}, \boldsymbol{\Sigma})$ and $\mathbf{x}, \boldsymbol{\mu}, \boldsymbol{\Sigma}$ are partitioned as:

$$\mathbf{x} = \begin{pmatrix} \mathbf{x}^{(1)} \\ \mathbf{x}^{(2)} \end{pmatrix}, \quad \boldsymbol{\mu} = \begin{pmatrix} \boldsymbol{\mu}^{(1)} \\ \boldsymbol{\mu}^{(2)} \end{pmatrix}, \quad \boldsymbol{\Sigma} = \begin{pmatrix} \boldsymbol{\Sigma}_{11} & \boldsymbol{\Sigma}_{12} \\ \boldsymbol{\Sigma}_{21} & \boldsymbol{\Sigma}_{22} \end{pmatrix},$$

where $\mathbf{x}^{(1)}$ and $\boldsymbol{\mu}^{(1)}: k \times 1$, $\boldsymbol{\Sigma}_{11}: k \times k$, then $\mathbf{x}^{(1)} \sim Mt_k(m, \boldsymbol{\mu}^{(1)}, \boldsymbol{\Sigma}_{11})$.

3. If $\mathbf{x} \overset{d}{=} R\mathbf{u}^{(n)} \sim Mt_n(m, \mathbf{0}, \mathbf{I}_n)$, then $R^2/m \sim \text{BeII}(n/2, m/2)$.

4. The c.f. of $\mathbf{x} \sim Mt_n(m, \boldsymbol{\mu}, \boldsymbol{\Sigma})$ is $\exp(i\mathbf{t}'\boldsymbol{\mu})\phi(\mathbf{t}'\boldsymbol{\Sigma}\mathbf{t})$ where ϕ is given by (3.30) with $N = \frac{1}{2}(n+m)$.

Sutradhar (1986) computes the integral (3.30) and obtains the c.f. of $\mathbf{y} \sim Mt_n(m, \boldsymbol{\mu}, \boldsymbol{\Sigma})$ as given in the following theorem.

Theorem 3.10

Let $\phi_{\mathbf{y}}(\mathbf{t}; m, \boldsymbol{\mu}, \boldsymbol{\Sigma})$ be the c.f. of $\mathbf{y} \sim Mt_n(m, \boldsymbol{\mu}, \boldsymbol{\Sigma})$. Then $\phi_{\mathbf{y}}(\mathbf{t}; m, \boldsymbol{\mu}, \boldsymbol{\Sigma})$ is given as in the following cases:

(a) For odd degrees of freedom m,

$$\phi_{\mathbf{y}}(\mathbf{t}; m, \boldsymbol{\mu}, \boldsymbol{\Sigma}) = \frac{\sqrt{\pi}\Gamma((m+1)/2)\exp(i\mathbf{t}'\boldsymbol{\mu} - \sqrt{\mathbf{t}'\boldsymbol{\Sigma}\mathbf{t}})}{2^{m-1}\Gamma(m/2)} \times \sum_{r=1}^{s}\left[\binom{2s-r-1}{s-r}\frac{(2\sqrt{m\mathbf{t}'\boldsymbol{\Sigma}\mathbf{t}})^{r-1}}{(r-1)!}\right],$$

where $s = (m+1)/2$.

(b) For even degrees of freedom m,

$$\phi_{\mathbf{y}}(\mathbf{t}; m, \boldsymbol{\mu}, \boldsymbol{\Sigma}) = \frac{(-1)^{s+1}\Gamma((m+1)/2)\exp(i\mathbf{t}'\boldsymbol{\mu})}{\sqrt{\pi}\prod_{j=1}^{s}(s+\frac{1}{2}-j)\Gamma(m/2)}\sum_{k=0}^{\infty} \times\left\{\left(\frac{m\mathbf{t}'\boldsymbol{\Sigma}\mathbf{t}}{4}\right)^{k}\frac{1}{(k!)^2}\left(\sum_{j=0}^{s-1}\prod_{\substack{l=0\\ l\neq j}}^{s-1}(k-l)\right) + \prod_{j=0}^{s-1}(k-j)\left[\log\frac{m(\mathbf{t}'\boldsymbol{\Sigma}\mathbf{t})}{4} - \psi(k+1)\right]\right\},$$

where $s = m/2$ and $\psi(k+1) = \Gamma'(k+1)/\Gamma(k+1)$ is the logarithmic derivative of the gamma function $\Gamma(k+1)$.

(c) For fractional degrees of freedom m,

$$\phi_{\mathbf{y}}(\mathbf{t}; m, \boldsymbol{\mu}, \boldsymbol{\Sigma}) = \frac{(-1)^{s}\Gamma((m+1)/2)\exp(i\mathbf{t}'\boldsymbol{\mu})(\pi/\sin\xi\pi)}{2^{\xi}\Gamma(m/2)\Gamma(\xi+\frac{1}{2})\prod_{j=1}^{s}\{(m+1)/2-j\}}m^{m/2-s} \times\sum_{k=0}^{\infty}\left[\left(\frac{m\mathbf{t}'\boldsymbol{\Sigma}\mathbf{t}}{4}\right)^{k}\frac{1}{(k!)}\left(\frac{2^{\xi}\prod_{j=0}^{s-1}(k-\xi-j)}{m^{\xi}\Gamma(k+1-\xi)} - \frac{(\mathbf{t}'\boldsymbol{\Sigma}\mathbf{t})^{\xi}\prod_{j=0}^{s-1}(k-j)}{2^{\xi}\Gamma(k+1+\xi)}\right)\right],$$

where $s = [(m+1)/2]$ is the integer part of $(m+1)/2$, $\xi = (m/2) - s$, and $0 < |\xi| < \frac{1}{2}$.

Details of the proof are given in Sutradhar's paper. (See also the correction in Sutradhar, 1988).

5. If $\mathbf{y}' = (Y_1, \ldots, Y_n) \sim Mt_n(m, \mathbf{0}, \mathbf{I}_n)$, then

$$E\left(\prod_{i=1}^{n} Y_i^{r_i}\right) = \begin{cases} m^l \dfrac{\Gamma((m-2l)/2)}{2^l\Gamma(m/2)} \displaystyle\prod_{i=1}^{n} \dfrac{(2l_i)!}{2^{l_i}(l_i)!}, & \text{if for } i = 1, \ldots, n, \\ & r_i = 2l_i \text{ are even} \\ 0, & \text{otherwise,} \end{cases}$$

where $r = r_1 + \cdots + r_n$, and $r = 2l$.

3.3.7 *The multivariate Cauchy distribution*

We now proceed to a brief discussion of the multivariate Cauchy distribution.

Definition 3.3
The distribution $Mt_n(1, \boldsymbol{\mu}, \boldsymbol{\Sigma})$ is called a **multivariate Cauchy** distribution and is denoted by $MC_n(\boldsymbol{\mu}, \boldsymbol{\Sigma})$.

Since the multivariate Cauchy distribution is a special case of the multivariate t-distribution, these two families share many properties in common. However, the well-known fact that the moments of the multivariate Cauchy distribution do not exist should be emphasized. In fact, this can be seen from (3.28).

If $\mathbf{x} \sim MC_n(\mathbf{0}, \mathbf{I}_n)$, its density is given by

$$\frac{\Gamma((n+1)/2)}{\pi^{(n+1)/2}}(1 + \mathbf{x}'\mathbf{x})^{-(n+1)/2}.$$

The marginal density of X_1 is

$$\frac{1}{\pi(1 + x_1^2)},$$

i.e., X_1 has a standard Cauchy distribution and its c.f. is

$$\exp(-|t|) = \exp(-\sqrt{(t^2)}),$$

where $\sqrt{u}$ denotes the positive square root of u. Thus, if $\mathbf{x} \sim MC_n(\boldsymbol{\mu}, \boldsymbol{\Sigma})$, then $\mathbf{x} \sim EC_n(\boldsymbol{\mu}, \boldsymbol{\Sigma}, \phi)$ with

$$\phi(u) = \exp(-\sqrt{u}).$$

Another multivariate version of the Cauchy distribution will be discussed in Chapter 7.

3.4 Symmetric multivariate Pearson Type II distributions

3.4.1 *Definition*

Definition 3.4

Let $\mathbf{x} \sim EC_n(\boldsymbol{\mu}, \boldsymbol{\Sigma}, g)$. If the density generator is of the form

$$g(u) = \frac{\Gamma(n/2 + m + 1)}{\Gamma(m+1)\pi^{n/2}}(1-u)^m, \quad 0 \leqslant u \leqslant 1, \quad m > -1, \tag{3.31}$$

then $\mathbf{x}$ is said to have a **symmetric multivariate Pearson Type II** distribution, and is denoted by $\mathbf{x} \sim MPII_n(\boldsymbol{\mu}, \boldsymbol{\Sigma})$.

Kotz (1975) defines a symmetric multivariate Pearson Type II distribution. M.E. Johnson (1987) presents a rather detailed discussion of this subclass of elliptical distributions. We shall only list here some basic results supplemented by concise calculations using somewhat different – possibly more convenient – notation.

The density of the generating variate r is

$$f(r) = \frac{2\Gamma(n/2 + m + 1)}{\Gamma(n/2)\Gamma(m+1)} r^{n-1}(1-r^2)^m, \quad 0 \leqslant r \leqslant 1, \tag{3.32}$$

and the density of $u = r^2$, the generating variate of $\mathbf{x}$, is thus

$$\frac{\Gamma(n/2 + m + 1)}{\Gamma(n/2)\Gamma(m+1)} u^{n/2-1}(1-u)^m$$

which implies that $r^2 \sim \text{Be}(n/2, m+1)$, a beta distribution.

3.4.2 *Moments and marginal distributions*

From (2.18) and the moments of beta distributions, we obtain for $\boldsymbol{\mu} = \mathbf{0}$ and $\boldsymbol{\Sigma} = \mathbf{I}_n$,

$$E\left(\prod_{i=1}^{n} X_i^{2s_i}\right) = \frac{\Gamma(n/2 + m + 1)}{\Gamma(n/2 + m + s + 1)\pi^{n/2}} \prod_{i=1}^{n} \Gamma(s_i + \tfrac{1}{2}), \tag{3.33}$$

where $s_1, \ldots, s_n$ are nonnegative integers and $s = s_1 + \cdots + s_n$. For $\mathbf{x} \sim MPII_n(\boldsymbol{\mu}, \boldsymbol{\Sigma})$, it is easy to verify that

$$E(\mathbf{x}) = \boldsymbol{\mu} \quad \text{and} \quad D(\mathbf{x}) = (n + 2m + 2)^{-1}\boldsymbol{\Sigma}.$$

Denote the p.d.f. of $(x_1, \ldots, x_k)$ of $\mathbf{x} \sim MPII_n(\mathbf{0}, \mathbf{I}_n)$ by $f_k(x_1^2 + \cdots + x_k^2)$.

Utilizing (2.23) we arrive at

$$f_k(u) = \frac{\pi^{(n-k)/2}}{\Gamma((n-k)/2)} \int_u^\infty (y-u)^{(n-k)/2-1} g(y)\,dy$$

$$= \frac{\Gamma(n/2+m+1)}{\Gamma((n-k)/2)\Gamma(m+1)\pi^{k/2}} \int_u^1 (y-u)^{(n-k)/2-1}(1-y)^m\,dy.$$

Making the transformation $w = (y-u)/(1-u)$ we obtain

$$f_k(u) = \frac{\Gamma(n/2+m+1)}{\Gamma((n-k)/2)\Gamma(m+1)\pi^{k/2}}(1-u)^{(n-k)/2+m}$$

$$\times \int_0^1 w^{(n-k)/2-1}(1-w)^m\,dw.$$

Thus

$$f_k(u) = \frac{\Gamma(n/2+m+1)}{\Gamma((n-k)/2+m+1)\pi^{k/2}}(1-u)^{(n-k)/2+m}. \tag{3.34}$$

In particular, the p.d.f. of X_1 is

$$f_1(x) = \frac{\Gamma(n/2+m+1)}{\Gamma((n-1)/2+m+1)\pi^{1/2}}(1-x^2)^{m+(n-1)/2}, \qquad |x| \leqslant 1, \tag{3.35}$$

which is also the p.d.f. of U_1, the first component of $\mathbf{u}^{(2m+n+2)}$ (cf. (3.9)).

3.4.3 *Characteristic function*

Let ϕ be the characteristic generator of $\mathbf{x} \sim MPII_n(\boldsymbol{\mu}, \boldsymbol{\Sigma})$. It is evident that $\phi(u^2) = E(\exp(iux_1))$. From (3.35) we obtain

$$E(e^{itx_1}) = C\int_{-1}^1 e^{itx}(1-x^2)^{m+(n-1)/2}\,dx$$

with

$$C = \frac{\Gamma(n/2+m+1)}{\Gamma((n-k)/2+m+1)\pi^{1/2}}.$$

Since

$$\int_{-1}^1 e^{itx}(1-x^2)^{m+(n-1)/2}\,dx$$

$$= 2\int_0^1 \cos tx(1-x^2)^{m+(n-1)/2}\,dx$$

$$= 2\sum_{j=0}^{\infty}\int_0^1 \frac{(tx)^{2j}}{(2j)!}(-1)^j(1-x^2)^{m+(n-1)/2}\,dx$$

$$= \sum_{j=0}^{\infty}(-1)^j\frac{t^{2j}}{(2j)!}\int_0^1 y^{j-1/2}(1-y)^{m+(n-1)/2}\,dy \qquad (\text{setting } y = x^2)$$

$$= \sum_{j=0}^{\infty}(-1)^j\frac{t^{2j}}{(2j)!}B\left(j+\frac{1}{2},\frac{n-1}{2}+m+1\right).$$

We have thus arrived at the following result.

Theorem 3.11
The characteristic generator of $MPII_n(\boldsymbol{\mu},\boldsymbol{\Sigma})$ is

$$\phi(u) = \frac{\Gamma(n/2+m+1)}{\Gamma((n-k)/2+m+1)}\sum_{j=0}^{\infty}(-1)^j\frac{u^j}{(2j)!}B\left(j+\frac{1}{2},\frac{n-1}{2}+m+1\right). \tag{3.36}$$

3.4.4 *Conditional densities*

Let $\mathbf{x} = \begin{pmatrix}\mathbf{x}^{(1)}\\ \mathbf{x}^{(2)}\end{pmatrix} \sim MPII_n(\mathbf{0},\mathbf{I}_n)$ with $\mathbf{x}^{(1)}$: $r\times 1$. By the definition of conditional density, we have

$$f(\mathbf{x}^{(1)}|\mathbf{x}^{(2)}) = \frac{\dfrac{\Gamma(n/2+m+1)}{\Gamma(m+1)\pi^{n/2}}(1-\mathbf{x}^{(1)\prime}\mathbf{x}^{(1)}-\mathbf{x}^{(2)\prime}\mathbf{x}^{(2)})^m}{\dfrac{\Gamma(n/2+m+1)}{\Gamma(k/2+m+1)\pi^{(n-k)/2}}(1-\mathbf{x}^{(2)\prime}\mathbf{x}^{(2)})^{m+k/2}}$$

$$= \frac{\Gamma(k/2+m+1)}{\Gamma(m+1)\pi^{k/2}}(1-\mathbf{x}^{(2)\prime}\mathbf{x}^{(2)})^{-k/2}\left(1-\frac{\mathbf{x}^{(1)\prime}\mathbf{x}^{(1)}}{1-\mathbf{x}^{(2)\prime}\mathbf{x}^{(2)}}\right).$$

Let $v = \mathbf{x}^{(2)\prime}\mathbf{x}^{(2)}$ and R_v be the generating variate of $(\mathbf{x}^{(1)}|\mathbf{x}^{(2)})$. Then the density of $u = R_v^2$ is

$$\frac{\Gamma(k/2+m+1)}{\Gamma(m+1)\Gamma(k/2)}(1-v)^{-k/2}u^{k/2-1}\left(1-\frac{u}{1-v}\right)^m,$$

$1-v\leqslant u\leqslant 1$, and $R_v^2/(1-v)\sim \text{Be}(k/2,m+1)$, which is analogous to $R^2\sim \text{Be}(n/2,m+1)$.

3.5 Some other subclasses of elliptically symmetric distributions

Definition 3.5

An $n \times 1$ random vector $\mathbf{x}$ is said to have a **symmetric multivariate Bessel** distribution if $\mathbf{x} \sim S_n(g)$ and the density generator g is of the form:

$$g(t) = C_n(t^{1/2}/\beta)^a K_a(t^{1/2}/\beta), \qquad a > -n/2, \quad \beta > 0,$$
$$C_n^{-1} = 2^{a+n-1}\pi^{n/2}\beta^n\Gamma(a+n/2), \tag{3.37}$$

where $K_a(\cdot)$ denotes the modified Bessel function of the third kind, i.e.,

$$K_a(z) = \frac{\pi}{2}\frac{I_{-a}(z) - I_a(z)}{\sin(a\pi)}, \qquad |\arg(z)| < \pi, \quad a = 0, \pm 1, \pm 2, \ldots, \tag{3.38}$$

where

$$I_a(z) = \sum_{k=0}^{\infty} \frac{1}{k!\Gamma(k+a+1)}\left(\frac{z}{2}\right)^{a+2k}, \qquad |z| < \infty, \quad |\arg(z)| < \pi.$$

If $\mathbf{x} \stackrel{d}{=} r\mathbf{u}^{(n)}$ has a symmetric multivariate Bessel distribution, then the density of r is

$$d_n \cdot r^{a+n-1}K_a(r/\beta),$$

where

$$d_n^{-1} = 2^{a+n-2}\beta^{a+n}\Gamma(\tfrac{1}{2}n)\Gamma(a+n/2).$$

In particular, setting $a = 0$ and $\beta = \sigma/\sqrt{2}$ in (3.37), we obtain

$$g(t) = C_n K_0(\sqrt{2t}/\sigma), \quad \sigma > 0, \tag{3.39}$$

where

$$C_n^{-1} = 2^{n/2-1}\pi^{n/2}\sigma^n\Gamma(n/2).$$

In this case, the distribution of $\mathbf{x}$ is called a **multivariate Laplace** distribution.

Jensen (1985) mentions two other elliptically symmetric distributions:

1. Elliptically symmetric logistic distribution. If $\mathbf{x} \sim EC_n(\boldsymbol{\mu}, \boldsymbol{\Sigma}, g)$ with

$$g(u) = C_n \exp(-u)/(1 + \exp(-u))^2, \quad u \geqslant 0$$

where

$$C_n = \frac{\pi^{n/2}}{\Gamma(n/2)}\int_0^{\infty} y^{n/2-1}\frac{e^{-y}}{(1+e^{-y})^2}\mathrm{d}y,$$

we say that $\mathbf{x}$ is distributed according to an **elliptically symmetric logistic** distribution and write $\mathbf{x} \sim ML_n(\boldsymbol{\mu}, \boldsymbol{\Sigma})$.

Several authors (Gumbel, 1961a; Malik and Abraham, 1973; Fang and Xu, 1989) have studied the multivariate logistic distribution using different definitions. We therefore use the designation 'elliptically symmetric logistic distribution' to single out our definition.

2. Multivariate symmetric stable laws. If the c.f. of a random vector $\mathbf{x}$ is of the form

$$\Phi(\mathbf{t}'\mathbf{t}) \equiv \exp(r(\mathbf{t}'\mathbf{t})^{\alpha/2}), \quad 0 < \alpha \leqslant 2$$

$r < 0$, we refer to $\mathbf{x}$ as having a **symmetric multivariate stable law**. This is a multivariate extension of the stable law. See e.g., Zolotarev (1981) for a detailed discussion on properties and estimation problems for these laws.

For $\alpha = 1$, the multivariate symmetric stable law reduces to a multivariate Cauchy distribution, discussed at the end of Section 3.3. The c.f. of X_1 is $\exp(r|t_1|^\alpha)$, which is the c.f. of a univariate stable law. Another multivariate version of a stable law was suggested by Cambanis, Keener and Simons (1983) and will be discussed in more detail in Chapter 7.

Cacoullos and Koutras (1984) and Koutras (1986) present a comprehensive discussion on a generalized noncentral χ^2-distribution and Fang (1984) concentrates in his paper on generalized noncentral t-, F- and Hotelling's T^2-distributions for the subclasses of symmetric multivariate Kotz type, Pearson Type VII and Bessel distributions. Mitchell and Krzanowski (1985) have studied the Rao distance, the Mahalanobis distance and the relation between them for elliptical distributions. Recently, Mitchell (1988) studied the statistical manifolds of univariate elliptical distributions. In particular, she has calculated many important characteristics (such as the information matrix, the α-connection, the Christoffel symbols of the first and second kind, the skewness tensor, etc.) for normal, Cauchy, multivariate t- or general elliptical distributions.

Problems

3.1 Let $U(-\frac{1}{2}, \frac{1}{2})$ denote the uniform distribution over $[-\frac{1}{2}, \frac{1}{2}]$. Show that there are no spherical distributions $S_n(\phi)$ having $U(-\frac{1}{2}, \frac{1}{2})$ as their marginal distributions.

3.2 Let $\mathbf{u}^{(m)}$ be uniformly distributed on the unit sphere in R^m. Show that there are no random vectors $\mathbf{u} \in R^{n-m}$, $0 < m < n$ such that

$$\begin{pmatrix} \mathbf{u}^{(m)} \\ \mathbf{u} \end{pmatrix} \stackrel{d}{=} \mathbf{u}^{(n)}.$$

3.3 Let $\mathbf{x}$ be distributed according to the uniform distribution in the unit sphere in R^n. Calculate the conditional distribution of $\mathbf{x}_{(1)}$ given $\mathbf{x}_{(2)}$, where $\mathbf{x}_{(1)}$: $m \times 1, 1 \leqslant m < n$, and $\mathbf{x} = (\mathbf{x}'_{(1)}, \mathbf{x}'_{(2)})'$.

3.4 Provide a detailed proof of (3.17).

3.5 Let $B \sim \mathrm{Be}(\alpha, \beta)$ and $B^* \sim \mathrm{BeII}(\alpha, \beta)$ (cf. Section 3.3). Derive the relationship between B and B^*.

3.6 Use the formula presented in Problem 2.3 to calculate the conditional density $\mathbf{x}_{(1)}$ given $\mathbf{x}_{(2)}$ for $\mathbf{x} = (\mathbf{x}'_{(1)}, \mathbf{x}'_{(2)})'$ distributed according to a symmetric multivariate Pearson Type II and a Pearson Type VII distribution.

3.7 Let $\mathbf{z} \sim S_n(g)$ and

$$W = \frac{\mathrm{d} \log g(\|\mathbf{z}\|^2)}{\mathrm{d}\|\mathbf{z}\|^2}.$$

Verify the following results:

(i) $E(\|\mathbf{z}\|^2 W) = -n/2$;

(ii) $E(\|\mathbf{z}\|^{2q} W^p) = (-1)^p 2^{q-p} \Gamma(q + \frac{1}{2}n)/\Gamma(\frac{1}{2}n)$, where $\mathbf{z} \sim N_n(\mathbf{0}, \mathbf{I}_n)$;

(iii) $$E(\|\mathbf{x}\|^{2q} W^p) = \frac{(-1)^p (m+n)^p \Gamma(\frac{1}{2}m + \frac{1}{2}n)}{2^p m^{p-q} \Gamma(\frac{1}{2}n)\Gamma(\frac{1}{2}m)} \times B\left(\frac{n}{2} + q, p - q + \frac{m}{2}\right),$$

if $\mathbf{z} \sim Mt_n(m, \mathbf{0}, \mathbf{I}_n)$

(cf. Mitchell, 1988).

3.8 Let $\{g(\mathbf{x}; \boldsymbol{\theta}), \boldsymbol{\theta} \in \Theta\}$ be a class of density functions with parameter vector $\boldsymbol{\theta} \in \Theta$. The corresponding Fisher information matrix is defined by $I(\boldsymbol{\theta}) = (I_{ij}(\boldsymbol{\theta}))$ with

$$I_{ij}(\boldsymbol{\theta}) = E_{\boldsymbol{\theta}}\left(\frac{\partial \log g(\mathbf{x}; \boldsymbol{\theta})}{\partial \theta_i} \frac{\partial \log g(\mathbf{x}; \boldsymbol{\theta})}{\partial \theta_j}\right).$$

Consider the density

$$g(\mathbf{x}; \boldsymbol{\mu}, \boldsymbol{\Sigma}) = |\boldsymbol{\Sigma}|^{-1/2} f((\mathbf{x} - \boldsymbol{\mu})'\boldsymbol{\Sigma}^{-1}(\mathbf{x} - \boldsymbol{\mu}))$$

as the above $g(\mathbf{x}; \boldsymbol{\theta})$ with $\boldsymbol{\theta} = (\boldsymbol{\mu}, \boldsymbol{\Sigma})$. Verify the following results:

(i) $$E\left(\frac{\partial \log g}{\partial \mu_i}\frac{\partial \log g}{\partial \mu_j}\right) = 4a_f\sigma^{ii},$$

where $\boldsymbol{\Sigma}^{-1} = (\sigma^{ij})$, $a_f = -E(\|\mathbf{z}\|^2 W)$, $\mathbf{z} = \boldsymbol{\Sigma}^{-1/2}(\mathbf{x} - \boldsymbol{\mu})$ and W is defined in Problem 3.7;

(ii) $$E\left(\frac{\partial \log g}{\partial \mu_i}\frac{\partial \log g}{\partial \sigma_{ij}}\right) = 0 \quad \text{with} \quad \boldsymbol{\Sigma} = (\sigma_{ij});$$

(iii) $$E\left(\frac{\partial \log g}{\partial \sigma_{kl}}\frac{\partial \log g}{\partial \sigma_{rs}}\right) = 2b_f \operatorname{tr}(\boldsymbol{\Sigma}^{-1}\mathbf{I}_{kl}\boldsymbol{\Sigma}^{-1}\mathbf{I}_{rs}) + \tfrac{1}{4}(4b_f - 1)\operatorname{tr}(\boldsymbol{\Sigma}^{-1}\mathbf{I}_{kl})\operatorname{tr}(\boldsymbol{\Sigma}^{-1}\mathbf{I}_{rs}),$$

where $b_f = E(\|\mathbf{z}\|^2 W)/n(n+2)$, $\mathbf{z}$ and W are defined in (i), and

$$\mathbf{I}_{ij} = \begin{cases} \mathbf{E}_{ii}, & \text{if } i = j \\ \mathbf{E}_{ij} + \mathbf{E}_{ji}, & \text{if } i \neq j \end{cases}$$

with $\mathbf{E}_{ij}$ being the $n \times n$ matrix with (i,j)th element 1 and 0 elsewhere (cf. Mitchell, 1988).

CHAPTER 4

Characterizations

In this chapter we shall focus our attention on characterizations of spherical distributions, uniformity and normality.

4.1 Some characterizations of spherical distributions

Several properties of spherical distributions which cannot apparently be extended to other multivariate distributions are discussed in this section. The main results are based on Eaton (1981, 1986), and on K.T. Fang's recent results.

Definition 4.1
Given an integer $n > 1$ and a random variable z, the distribution of a random vector $\mathbf{x} \in R^n$ is called an n-dimensional version of z if there exists a function c on R^n to $[0, \infty)$ such that

(i) $c(\mathbf{a}) > 0 \quad \text{if } \mathbf{a} \neq \mathbf{0}$;

(ii) $\mathbf{a}'\mathbf{x} \overset{d}{=} c(\mathbf{a})z, \quad \mathbf{a} \in R^n$.

The distribution of $\mathbf{x}$ is called an ***n*-dimensional spherical version** of z or ***n*-dimensional isotropic version** of z if it is an n-dimensional version of z and $\mathbf{x}$ is spherical.

Definition 4.2
The set of all distributions in $S_1^+(\phi)$, $\phi \in \Phi_1$, which possess n-dimensional versions will be denoted by G_n, $n \geqslant 2$.

Definition 4.3
The class of m-marginals of elements of $S_{m+n}^+(\phi)$, $\phi \in \Phi_{m+n}$ is denoted by $D(m, n)$.

It is evident that $D(1, n-1) \subseteq G_n$, $n = 2, \ldots$. Our first result identifies those distributions in G_n which have a finite variance.

Theorem 4.1
If the distribution of z belongs to G_n and $\mathrm{Var}(z) < +\infty$, then the distribution of z belongs to $D(1, n-1)$ and every n-dimensional version of z is given by $\mathbf{Ax}_0$, where $\mathbf{x}_0$ is an n-dimensional spherical version of z and $\mathbf{A}$ is an $n \times n$ nonsingular matrix.

PROOF Let $\mathbf{x}$ be an n-dimensional version of z with $c(\cdot)$ satisfying Definition 4.1. Since $\mathrm{Var}(z) = \sigma^2 < \infty$, it follows that $\mathbf{x}$ has a covariance matrix, say $\mathbf{\Sigma} = \mathrm{Cov}(\mathbf{x})$. The relation $\mathbf{a}'\mathbf{x} \stackrel{d}{=} c(\mathbf{a})\mathbf{z}$ implies $c(\mathbf{a}) = (\mathbf{a}'\mathbf{\Sigma a})^{1/2}/\sigma$, $\mathbf{a} \in R^n$, which shows that $\mathbf{\Sigma}$ is positive definite since $c(\mathbf{a}) > 0$ for $\mathbf{a} \neq \mathbf{0}$. Set $\mathbf{x}_0 = \sigma\mathbf{\Sigma}^{-1/2}\mathbf{x}$ where $\mathbf{\Sigma}^{-1/2}$ denotes the inverse of the positive definite square root of $\mathbf{\Sigma}$. For $\mathbf{a} \in R^n$,

$$\mathbf{a}'\mathbf{x}_0 \stackrel{d}{=} (\sigma\mathbf{\Sigma}^{-1/2}\mathbf{a})'\mathbf{x} \stackrel{d}{=} c(\sigma\mathbf{\Sigma}^{-1/2}\mathbf{a})z \stackrel{d}{=} \|\mathbf{a}\| z,$$

which implies that $\mathbf{x}_0$ is an n-dimensional spherical version of z. Thus, the distribution of z belongs to $D(1, n-1)$ and, of course, the distribution of $\mathbf{x}_0$ is unique since the distribution of $\mathbf{x}_0$ belongs to $S_n^+(\phi)$ and $\mathbf{a}'\mathbf{x}_0 \stackrel{d}{=} \|\mathbf{a}\| z$. Setting $\mathbf{A} = (1/\sigma)\mathbf{\Sigma}^{1/2}$ we complete the proof. □

This theorem shows us that if z has a finite variance, each n-dimensional version of z must have an elliptical distribution. In Chapter 7, we will continue the discussion of the same problem for α-symmetric multivariate distributions. Utilizing a different approach the following theorem will give us another characterization of spherical distributions.

Theorem 4.2
An $n \times 1$ random vector $\mathbf{x}$ with a finite mean vector has a spherical distribution if and only if for any perpendicular pair of vectors $\mathbf{u} \neq 0$ and $\mathbf{v} \neq \mathbf{0}$ (i.e., $\mathbf{u}'\mathbf{v} = 0$),

$$E(\mathbf{u}'\mathbf{x} \mid \mathbf{v}'\mathbf{x}) = 0. \tag{4.1}$$

PROOF Since $\mathbf{x}$ has a mean vector, the gradient of $g(\mathbf{t})$, the c.f. of $\mathbf{x}$, exists and is given by

$$\nabla g(\mathbf{t}) = iE(\mathbf{x}\exp(i\mathbf{t}'\mathbf{x})). \tag{4.2}$$

Simple conditioning arguments show that (4.1) implies

$$E\{\mathbf{u}'\mathbf{x}\exp(i\mathbf{v}'\mathbf{x})\} = 0, \tag{4.3}$$

which in turn is equivalent to

$$\mathbf{u}'\nabla g(\mathbf{v}) = 0. \tag{4.4}$$

Now assume that (4.1) holds for orthogonal pairs. We shall use (4.4) to verify that

$$g(\mathbf{t}) = g(\mathbf{\Gamma t}) \quad \text{for any } \mathbf{\Gamma} \in O(n), \tag{4.5}$$

provided $\mathbf{t} \neq \mathbf{0}$. Since $\mathbf{\Gamma t}$ must lie on the sphere with centre $\mathbf{0}$ and radius $r = \|\mathbf{t}\|$, there exists on this sphere a smooth curve passing through $\mathbf{t}$ and $\mathbf{\Gamma t}$. This curve can be represented by a continuously differentiable function $\mathbf{c}(\alpha)$ from the interval $(0, 1)$ to the surface of the sphere such that $\mathbf{c}(\alpha_1) = \mathbf{t}$ and $\mathbf{c}(\alpha_2) = \mathbf{\Gamma t}$ for some $\alpha_1, \alpha_2 \in (0, 1)$. Since $\|\mathbf{c}(\alpha)\| = r$ for $\alpha \in (0, 1)$, we have

$$(\dot{\mathbf{c}}(\alpha))'\mathbf{c}(\alpha) = 0 \quad \text{for all } \alpha \in (0, 1), \tag{4.6}$$

where $\dot{\mathbf{c}}(\alpha)$ is the vector of derivatives of components of $\mathbf{c}$. Using (4.4) for the orthogonal pair given in (4.6) we have

$$\frac{\mathrm{d}}{\mathrm{d}\alpha} g(\mathbf{c}(\alpha)) = (\dot{\mathbf{c}}(\alpha))'\nabla g(\mathbf{c}(\alpha)) = 0,$$

for $\alpha \in (0, 1)$. Thus $g(\mathbf{c}(\alpha))$ is constant in α, so that (4.5) is valid and hence $\mathbf{x}$ has a spherical distribution.

Conversely, let $\mathbf{x} \sim S_n(\phi)$ for some ϕ and $E(\|\mathbf{x}\|) < \infty$. From the assumption of orthogonality $\mathbf{u}'\mathbf{v} = 0$, there exists a $\mathbf{\Gamma} \in O(n)$ such that

$$\mathbf{\Gamma u} = (\|\mathbf{u}\|, 0, \ldots, 0)' \quad \text{and} \quad \mathbf{\Gamma v} = (0, \|\mathbf{v}\|, 0, \ldots, 0)'.$$

Thus

$$\begin{aligned}(\mathbf{u}'\mathbf{x} \,|\, \mathbf{v}'\mathbf{x}) &\overset{d}{=} (\mathbf{u}'\mathbf{\Gamma}'\mathbf{x} \,|\, \mathbf{v}'\mathbf{\Gamma}'\mathbf{x}) \\ &= (\|\mathbf{u}\| X_1 \,|\, \|\mathbf{v}\| X_2) \sim S_1(\phi^*)\end{aligned}$$

for some ϕ^* (cf. (2.28)) and

$$E(\mathbf{u}'\mathbf{x} \,|\, \mathbf{v}'\mathbf{x}) = E(\|\mathbf{u}\| X_1 \,|\, \|\mathbf{v}\| X_2) = 0.$$

The proof is completed. □

Theorem 4.3

A necessary and sufficient condition for $\mathbf{x}$ to have a spherical distribution is that for any pair of perpendicular vectors $\mathbf{v} \neq \mathbf{0}$ and

$\mathbf{u} \neq \mathbf{0}$,

$$(\mathbf{u}'\mathbf{x} \mid \mathbf{v}'\mathbf{x}) \stackrel{d}{=} (-\mathbf{u}'\mathbf{x} \mid \mathbf{v}'\mathbf{x}). \tag{4.7}$$

PROOF Let $g(\cdot)$ be the c.f. of $\mathbf{x}$. It is easy to show that (4.7) is equivalent to

$$g(a\mathbf{u} + b\mathbf{v}) = g(-a\mathbf{u} + b\mathbf{v}) \tag{4.8}$$

for all $a, b \in R$ and $\mathbf{u}$ which are perpendicular to $\mathbf{v} \neq \mathbf{0}$. Continuity (4.8) holds whenever $\mathbf{v} = \mathbf{0}$. To show (4.5), consider $\mathbf{t}$ and $\mathbf{\Gamma t}$, and set $\mathbf{v} = \frac{1}{2}(\mathbf{\Gamma t} + \mathbf{t})$. With $\mathbf{u} = \frac{1}{2}(\mathbf{\Gamma t} - \mathbf{t}), \mathbf{u}'\mathbf{v} = 0$ and

$$\mathbf{u} + \mathbf{v} = \mathbf{\Gamma t}, \quad -\mathbf{u} + \mathbf{v} = \mathbf{t}.$$

Thus, (4.8) yields (4.5) and hence $\mathbf{x}$ is spherical. The converse is trivial. □

Some applications of Theorems 4.2 and 4.3 in the linear model are given by Eaton, 1986.

Since the spherical distributions possess the stochastic structure $\mathbf{x} \stackrel{d}{=} r\mathbf{u}^{(n)}$, a more general family can be considered. Let $\mathbf{y}$ be an $n \times 1$ random vector and

$$\mathscr{F}(\mathbf{y}) = \{\mathbf{x} : \mathbf{x} \stackrel{d}{=} r\mathbf{y}, r \geqslant 0 \text{ is independent of } \mathbf{y}\}. \tag{4.9}$$

The following are some basic properties of $\mathscr{F}(\mathbf{y})$:

1. If $\mathbf{z} \in \mathscr{F}(\mathbf{y})$, then $\mathscr{F}(\mathbf{z}) \subset \mathscr{F}(\mathbf{y})$.
2. Let $\mathbf{y}$ and $\mathbf{z}$ be two n-dimensional random vectors satisfying $\mathbf{y} = c\mathbf{z}$, $c \neq 0$. Then $\mathscr{F}(\mathbf{y}) = \mathscr{F}(\mathbf{z})$.

Two n-dimensional random vectors $\mathbf{y}$ and $\mathbf{z}$ are said to be **equivalent** if $\mathscr{F}(\mathbf{y}) = \mathscr{F}(\mathbf{z})$ and will be denoted by $\mathbf{y} \sim \mathbf{z}$. Using this terminology the above assertion can be re-expressed as follows: if $\mathbf{y} = c\mathbf{z}$ then $\mathbf{y} \sim \mathbf{z}$.

Example 4.1

Let $\mathbf{y} \sim N_n(\mathbf{0}, \mathbf{I}_n)$ and $\mathbf{z} \sim Mt_n(m, \mathbf{0}, \mathbf{I}_n)$, a multivariate t-distribution. Since

$$\mathbf{z} \stackrel{d}{=} \frac{1}{S}\mathbf{y},$$

where $S \sim \chi_m$ is independent of $\mathbf{y}$, we have $\mathbf{z} \in \mathscr{F}(\mathbf{y})$. Since $S^{-1} \not\equiv$ constant, $\mathscr{F}(\mathbf{z}) \subset \mathscr{F}(\mathbf{y})$ and $\mathscr{F}(\mathbf{z})$ is a proper subclass of $\mathscr{F}(\mathbf{y})$, the latter being the class of mixtures of normal distributions.

Denote $S_1 = \mathscr{F}(\mathbf{y})$, $S_2 = \mathscr{F}(\mathbf{z})$ and $S = \mathscr{F}(\mathbf{u}^{(n)})$, where $\mathbf{u}^{(n)}$ is uniformly distributed on the unit sphere. Evidently $\mathbf{y} \in S_2 \subset S_1 \subset S$. The following theorem shows us that the class S is the largest in a certain sense.

Theorem 4.4
Under the same notation as above the following are valid:

(1) there is no n-dimensional random vector $\mathbf{v}$ such that $S \subset \mathscr{F}(\mathbf{v})$ unless $\mathbf{v} \sim \mathbf{u}^{(n)}$;
(2) if $\mathbf{x} \in S$ and $\mathbf{x} \in \mathscr{F}(\mathbf{u})$ for some $\mathbf{u}$, and $P(\mathbf{X} = \mathbf{0}) = 0$, then $\mathscr{F}(\mathbf{u}) \in S$, or $\mathbf{u}$ is spherical.

PROOF Assume that there exists a $\mathbf{v}$ such that $S \subset \mathscr{F}(\mathbf{v})$. This implies the existence of $r \geqslant 0$ such that $\mathbf{u}^{(n)} \stackrel{d}{=} r\mathbf{v}$. Thus

$$1 = \mathbf{u}^{(n)\prime}\mathbf{u}^{(n)} = r^2\mathbf{v}'\mathbf{v}.$$

Since r is independent of $\mathbf{v}$, r^2 is independent of $\mathbf{v}'\mathbf{v}$; the fact that $r^2\mathbf{v}'\mathbf{v} = 1$ implies $r =$ constant, hence assertion 1 follows. To prove 2, assume that there exist $r > 0$ and $r^* > 0$ such that $\mathbf{X} \stackrel{d}{=} r\mathbf{u}$ and $\mathbf{X} \stackrel{d}{=} r^*\mathbf{u}^{(n)}$. We have

$$\mathbf{u}/\|\mathbf{u}\| = r\mathbf{u}/r\|\mathbf{u}\| = \mathbf{x}/\|\mathbf{x}\| \stackrel{d}{=} \mathbf{u}^{(n)},$$

which is independent of $\|\mathbf{x}\| = r\|\mathbf{u}\|$, and hence independent of $\|\mathbf{u}\|$ because r and $\|\mathbf{u}\|$ are independent. Therefore $\mathbf{u}$ has a spherical distribution. □

The above discussion is based on the results of Fang and Bentler, 1989. Misiewicz (1984) presents a discussion of the elliptical distribution on a Banach space and some characterizations on an infinite-dimensional Banach space.

4.2 Characterizations of uniformity

In this section several characterizations of uniformity are discussed.

Let $S = S_n$ be the unit sphere in R^n and $C(\mathbf{x}, r)$ be a ball with centre $\mathbf{x}$ and radius r, i.e.,

$$C(\mathbf{x}, r) = \{\mathbf{y}: d(\mathbf{y}, \mathbf{x}) \leqslant r, \mathbf{y} \in R^n\}, \tag{4.10}$$

where $d(\mathbf{x}, \mathbf{y}) = [(\mathbf{x} - \mathbf{y})'(\mathbf{x} - \mathbf{y})]^{1/2}$.

If $\mathbf{x} \stackrel{d}{=} \mathbf{u}^{(n)}$, the uniform distribution on S, it can be verifiied that

$$P(\mathbf{x} \in C(\mathbf{y}, r)) = C_r \quad \text{for each} \quad \mathbf{y} \in S \quad \text{and} \quad r > 0, \tag{4.11}$$

where C_r is a constant depending on r only. The following theorem shows us that this fact can be used for a characterization of uniformity.

Theorem 4.5

Let $\mathbf{x}$ be an $n \times 1$ random vector defined on S. If (4.11) holds for each $\mathbf{y} \in S$ and $r > 0$, then $\mathbf{x} = \mathbf{u}^{(n)}$.

PROOF Firstly we shall verify the following fact: if μ is a measure on S and

$$\mu(C(\mathbf{y}, r)) = e_r \quad \text{for each} \quad y \in S \quad \text{and} \quad r > 0, \tag{4.12}$$

then $e_r = C_r$ for each $r > 0$.

Let f be any bounded continuous function from S to R and $P(\cdot)$ be the probability measure generated by $\mathbf{x}$. By the Fubini theorem we have, for each $r > 0$,

$$\begin{aligned}\int_S f(\mathbf{y})\,\mathrm{d}p(\mathbf{y}) &= \int_S f(\mathbf{y}) \left\{ \frac{1}{e_r} \int_{C(\mathbf{y},r)} \mathrm{d}\mu(\mathbf{x}) \right\} \mathrm{d}p(\mathbf{y}) \\ &= \frac{1}{e_r} \int_S \left\{ \int_{C(\mathbf{y},r)} f(\mathbf{y})\,\mathrm{d}p(\mathbf{y}) \right\} \mathrm{d}\mu(\mathbf{x}).\end{aligned}$$

Choosing $f \equiv 1$ implies that $e_r = C_r$.

For each $\mathbf{x} \in S$,

$$\left| \frac{1}{e_r} \int_{C(\mathbf{y},r)} f(\mathbf{y})\,\mathrm{d}p(\mathbf{y}) - f(\mathbf{x}) \right| \leqslant \sup\{|f(\mathbf{y}) - f(\mathbf{x})| : \mathbf{y} \in C(\mathbf{x}, r)\}$$

$\to 0$ as $r \to 0$. The bounded convergence theorem now implies that for each bounded continuous function f we have

$$\int_S f(\mathbf{x})\,\mathrm{d}\mu(\mathbf{x}) = \int_S f(x)\,\mathrm{d}p(\mathbf{x}),$$

i.e., $\mu \equiv p$. In particular, choosing μ to be the measure generated by $\mathbf{u}^{(n)}$ satisfying (4.12), we have $p =$ the measure generated by $\mathbf{u}^{(n)}$, i.e. $\mathbf{x} \stackrel{d}{=} \mathbf{u}^{(n)}$. □

Let $\mathbf{x}$ and $\mathbf{y}$ be two $n \times 1$ random vectors defined on S, and $\mathbf{x}$ and $\mathbf{y}$ be independent. Let S be the support of $\mathbf{x}$, i.e.,

$$p(\mathbf{x} \in C(\mathbf{v}, r)) > 0, \quad \text{for each} \quad \mathbf{v} \in S, r > 0. \tag{4.13}$$

If $\mathbf{y} \stackrel{d}{=} \mathbf{u}^{(n)}$, then

$$p(d(\mathbf{x}, \mathbf{y}) \leqslant r \mid \mathbf{x} = \mathbf{x}_0) = p(\mathbf{y} \in C(\mathbf{x}_0, r)) = C_r,$$

which is independent of $\mathbf{x}$, or $d(\mathbf{x}, \mathbf{y})$ is independent of $\mathbf{x}$. This observation yields another characterization of uniformity.

Theorem 4.6
Let $\mathbf{x}$ and $\mathbf{y}$ be two $n \times 1$ independent random vectors defined on S. Suppose that $\mathbf{x}$ has a support S. If distance $d(\mathbf{x}, \mathbf{y})$ is independent of $\mathbf{x}$, then $\mathbf{y} \stackrel{d}{=} \mathbf{u}^{(n)}$.

PROOF By the triangle inequality for $\mathbf{v} \in S$ and $0 < \varepsilon < r$ we have

$$\begin{aligned} &P(\mathbf{x} \in C(\mathbf{v}, \varepsilon), \quad \mathbf{y} \in C(\mathbf{v}, r - \varepsilon)) \\ &\leqslant P(\mathbf{x} \in C(\mathbf{v}, \varepsilon), \quad d(\mathbf{x}, \mathbf{y}) \leqslant r) \\ &\leqslant P(\mathbf{x} \in C(\mathbf{v}, \varepsilon), \quad \mathbf{y} \in C(\mathbf{v}, r + \varepsilon)). \end{aligned}$$

Since S is the support of $\mathbf{x}$, we can divide the two sides of the above inequality by $P(\mathbf{x} \in V(\mathbf{v}, \varepsilon))$ and use the independence of $\mathbf{x}$ and $\mathbf{y}$ and that of $\mathbf{x}$ and $d(\mathbf{x}, \mathbf{y})$ to obtain

$$P(\mathbf{y} \in C(\mathbf{v}, r - \varepsilon)) \leqslant P(d(\mathbf{x}, \mathbf{y}) \leqslant r) \leqslant P(\mathbf{y} \in C(\mathbf{v}, r + \varepsilon)).$$

Letting $\varepsilon \to 0$, we find that $P(d(\mathbf{x}, \mathbf{y}) \leqslant r)$ lies between $P(d(\mathbf{y}, \mathbf{v}) < r)$ and $P(d(\mathbf{y}, \mathbf{v}) \leqslant r)$. Since $\{d(\mathbf{y}, \mathbf{v}) = r\}$ are disjoint events for different r, $P(d(\mathbf{y}, \mathbf{v}) = r) = 0$ except for at most countably many r. The last two observations imply that for fixed $\mathbf{v}$,

$$P(\mathbf{y} \in C(\mathbf{v}, r)) = P(d(\mathbf{x}, \mathbf{y}) \leqslant r), \tag{4.14}$$

except for at most countably many r. By continuity of the two sides of (4.14) in r, (4.14) holds for each $r > 0$. The right-hand side of (4.14) is thus independent of $\mathbf{v}$ which implies $\mathbf{y} \stackrel{d}{=} \mathbf{u}^{(n)}$ by Theorem 4.5. □

Theorem 4.7
Let $\mathbf{x}, \mathbf{y}$ and $\mathbf{z}$ be $n \times 1$ mutually independent random vectors defined on S. If S is the support of $\mathbf{x}$, and $d(\mathbf{x}, \mathbf{z})$ and $d(\mathbf{x}, \mathbf{y})$ are independent, then $\mathbf{y} \stackrel{d}{=} \mathbf{z} \stackrel{d}{=} \mathbf{u}^{(n)}$.

PROOF By a method similar to one in the proof of Theorem 4.6, for $0 < \varepsilon < r$, we have:

$$\begin{aligned} &P(\mathbf{x} \in C(\mathbf{z}, \varepsilon), \quad \mathbf{y} \in C(\mathbf{z}, r - \varepsilon)) \\ &\leqslant P(\mathbf{x} \in C(\mathbf{z}, \varepsilon), \quad d(\mathbf{x}, \mathbf{y}) \leqslant r) \\ &= P(d(\mathbf{x}, \mathbf{z}) \leqslant \varepsilon, \quad d(\mathbf{x}, \mathbf{y}) \leqslant r) \\ &\leqslant P(\mathbf{x} \in C(\mathbf{z}, \varepsilon), \quad \mathbf{y} \in C(\mathbf{z}, r + \varepsilon)). \end{aligned}$$

Since S is the support of $\mathbf{x}$, using the independence among $\mathbf{x}$, $\mathbf{y}$ and $\mathbf{z}$ and between $d(\mathbf{x}, \mathbf{y})$ and $d(\mathbf{x}, \mathbf{z})$, we have

$$P(\mathbf{y} \in C(\mathbf{z}, r - \varepsilon)) \leqslant P(d(\mathbf{x}, \mathbf{y}) \leqslant r) \leqslant P(\mathbf{y} \in C(\mathbf{z}, r + \varepsilon)).$$

Let $\varepsilon \to 0$, then we obtain by continuity

$$P(d(\mathbf{y}, \mathbf{z}) < r) \leqslant P(d(\mathbf{x}, \mathbf{y}) \leqslant r) \leqslant P(d(\mathbf{y}, \mathbf{z}) \leqslant r)$$

and

$$P(d(\mathbf{x}, \mathbf{y}) \leqslant r) = P(\mathbf{y} \in C(\mathbf{z}, r)) \quad \text{for each} \quad r > 0,$$

which implies that $d(\mathbf{x}, \mathbf{y})$ is independent of $\mathbf{x}$. By Theorem 4.6, $\mathbf{y} \stackrel{d}{=} \mathbf{u}^{(n)}$. Similarly $\mathbf{z} \stackrel{d}{=} \mathbf{u}^{(n)}$. □

The above three theorems are essentially due to Brown, Cartwright and Eagleson (1986).

In Section 2.7 we showed that the probability of a cone is invariant in the class of $EC_n(\mathbf{0}, \mathbf{\Sigma}, \phi)$ distributions without an atom at $\mathbf{0}$. This property can be used for characterization of the uniform distribution and the elliptical distribution.

Theorem 4.8
Let $\mathbf{y} \sim N_n(\mathbf{0}, \mathbf{I}_n)$ and $\mathbf{x}$ be an $n \times 1$ random vector with $\mathbf{x}'\mathbf{x} = 1$ having a p.d.f on the S. If the probability that $\mathbf{x} \in C$ equals the corresponding probability $\mathbf{y} \in C$ for each cone (2.63), then $\mathbf{x} \stackrel{d}{=} \mathbf{u}^{(n)}$.

PROOF By the assumption $\mathbf{x}$ can be expressed in terms of $(\cos\theta_1, \sin\theta_1 \cos\theta_2, \ldots, (\prod_{k=1}^{n-2} \sin\theta_k)\cos\theta_{n-1}, \prod_{k=1}^{n-1} \sin\theta_k)$ and $(\theta_1, \ldots, \theta_{n-1})$

has a density $g(\theta_1,\ldots,\theta_{n-1})$. Thus

$$P(\mathbf{x}\in C)=\int_C g(\theta_1,\ldots,\theta_{n-1})\,\mathrm{d}\theta_1\ldots\mathrm{d}\theta_{n-1}.$$

Taking $\alpha_k^1=0, \alpha_k^2=\alpha_k$ in (2.63), $k=1,\ldots,n-1$, we have

$$P(\mathbf{x}\in C)=\int_0^{\alpha_1}\cdots\int_0^{\alpha_{n-1}} g(\theta_1,\ldots,\theta_{n-1})\,\mathrm{d}\theta_1\ldots\mathrm{d}\theta_{n-1}.$$

On the other hand, it is easy to verify (cf. Section 2.7) that

$$P(\mathbf{y}\in C)=\Gamma(n/2)/(2\pi^{n/2})\int_0^{\alpha_1}\cdots\int_0^{\alpha_{n-1}}\prod_{k=1}^{n-2}(\sin^{n-k-1}\theta_k\,\mathrm{d}\theta_k).$$

The assumption $P(\mathbf{x}\in C)=P(\mathbf{y}\in C)$ for each cone implies that

$$g(\alpha_1,\ldots,\alpha_{n-1})=(\Gamma(n/2)/2\pi^{n/2})\prod_{k=1}^{n-2}\sin^{n-k-1}\alpha_k,$$

i.e., $\theta_1,\ldots,\theta_{n-1}$ are independently distributed with the following p.d.f.s:

$$h(\theta_k)=\frac{1}{B(\frac{1}{2},(n-k)/2)}\sin^{n-k-1}\theta_k,\quad 0\leqslant\theta_k<\pi,\quad k=1,\ldots,n-2$$

$$h(\theta_{n-1})=\frac{1}{2\pi},\qquad 0\leqslant\theta_{n-1}<2\pi. \tag{4.15}$$

The assertion $\mathbf{x}\overset{d}{=}\mathbf{u}^{(n)}$ now follows from Theorem 2.11. □

Theorem 4.9
Let $\mathbf{x}$ be an $n\times 1$ random vector with $P(\mathbf{x}=\mathbf{0})=0$. Suppose that $\|\mathbf{x}\|$ and $\mathbf{x}/\|\mathbf{x}\|$ are independent and $\mathbf{x}$ has a density function. If

$$P(\mathbf{x}\in C)=P(\mathbf{y}\in C)\quad\text{for each cone (2.63)}, \tag{4.16}$$

where $\mathbf{y}\sim N_n(\mathbf{0},I_n)$, then $\mathbf{x}$ is spherical.

PROOF Let r, $\theta_1,\ldots,\theta_{n-1}$ be the spherical coordinates of $\mathbf{x}$ (cf. (2.20)). The independence between $\|\mathbf{x}\|$ and $\mathbf{x}/\|\mathbf{x}\|$ implies that r and $\{\theta_1,\ldots,\theta_{n-1}\}$ are independent. Hence the joint density of r, $\theta_1,\ldots,\theta_{n-1}$ has the form

$$f(r,\theta_1,\ldots,\theta_{n-1})=g(r)h(\theta_1,\ldots,\theta_{n-1}).$$

The assumption (4.16) implies that

$$\begin{aligned}\int_C & h(\theta_1,\ldots,\theta_{n-1})\,d\theta_1\ldots d\theta_{n-1}\\ &= \int_C h(\theta_1,\ldots,\theta_{n-1})\,d\theta_1\ldots d\theta_{n-1}\int_0^\infty g(r)\,dr\\ &= \Gamma(n/2)/(2\pi^{n/2})\int_C\left[\prod_{k=1}^{n-2}\sin^{n-k-1}\theta_k\,d\theta_k\right].\end{aligned}$$

By an argument similar to that given in the proof of Theorem 4.8, we can show that $\theta_1,\ldots,\theta_{n-1}$ are independently distributed with p.d.f.s (4.15), i.e., $\mathbf{x}/\|\mathbf{x}\| \stackrel{d}{=} \mathbf{u}^{(n)}$. The assertion now follows from the independence of $\|\mathbf{x}\|$ and $\mathbf{x}/\|\mathbf{x}\|$. □

Corollary
Let $\mathbf{y} \sim N_n(\mathbf{0},\boldsymbol{\Sigma}), \boldsymbol{\Sigma} > \mathbf{0}$, and $\mathbf{x}$ be an $n\times 1$ random vector with $P(\mathbf{x}=\mathbf{0})=0$. Suppose that $\mathbf{x}'\boldsymbol{\Sigma}^{-1}\mathbf{x}$ and $\boldsymbol{\Sigma}^{-1/2}\mathbf{x}/\mathbf{x}'\boldsymbol{\Sigma}^{-1}\mathbf{x}$ are independent and that $\mathbf{x}$ has a density function. If (4.16) holds for each cone (2.63), then $\mathbf{x} \sim EC_n^+(\mathbf{0},\boldsymbol{\Sigma},\phi)$ for some $\phi\in\Phi_n$.

The last two theorems are due to Wang (1987).

4.3 Characterization of normality

The normal distributions are of course members of the class of elliptically symmetric distributions. In this section we shall study several characterizations of normality within this class. The main results in this section are based on Kelker (1970), Cambanis, Huang and Simons (1981), Anderson and Fang (1982, 1987), and Khatri and Mukerjee (1987).

Theorem 4.10
Assume that $\mathbf{x} \sim EC_n(\boldsymbol{\mu},\boldsymbol{\Sigma},\phi)$. Then any marginal distribution is normal if and only if $\mathbf{x}$ has a multinormal distribution.

PROOF Assume $\mathbf{x}^{(1)}$ where $\mathbf{x} = \begin{pmatrix}\mathbf{x}^{(1)}\\ \mathbf{x}^{(2)}\end{pmatrix}$ has a normal distribution. Since the c.f. of $\mathbf{x}^{(1)}$ is $\phi(\mathbf{t}'\mathbf{t})$ where $\mathbf{t}$ is of the same dimension as $\mathbf{x}^{(1)}$, we have from the normality of $\mathbf{x}^{(1)}$ that $\phi(u)=\exp(-\frac{1}{2}u)$. Thus $\mathbf{x}$ is normally distributed. The implication in the other direction is obvious. □

Note that without the above assumption on $\mathbf{x}$ the theorem is generally not valid.

Theorem 4.11
Assume $\mathbf{x} \sim EC_n(\boldsymbol{\mu}, \boldsymbol{\Sigma}, \phi)$ with $\boldsymbol{\Sigma} = \text{diag}(\sigma_{11}, \ldots, \sigma_{pp})$. Then the following statements are equivalent:

(a) $\mathbf{x}$ is normally distributed;
(b) the components of $\mathbf{x}$ are independent;
(c) x_i and $x_j (1 \leqslant i < j \leqslant n)$ are independent.

PROOF The implications (a)$\Rightarrow$(b)$\Rightarrow$(c) are trivial. To prove that (c)$\Rightarrow$(a), note that by the assumption

$$\phi(t_i^2 \sigma_{ii} + t_j^2 \sigma_{jj}) = \phi(t_i^2 \sigma_{ii}) \phi(t_j^2 \sigma_{jj}), \qquad (\text{setting } u_k = t_k \sigma_{kk}^{1/2}, k = i, j)$$

then

$$\phi(u_i + u_j) = \phi(u_i) \phi(u_j).$$

This equation, known as Hamel's equation, has a solution $\phi(u) = \exp(-au)$ for some constant a (although $\phi(u) = 1$ is also a solution of the equation, the corresponding distribution is singular). Since $\phi(u)$ is a c.f., the constant a must be positive, which completes the proof. □

Theorem 4.12
Assume $\mathbf{x} = \begin{pmatrix} \mathbf{x}^{(1)} \\ \mathbf{x}^{(2)} \end{pmatrix} \sim EC_n(\boldsymbol{\mu}, \boldsymbol{\Sigma}, \phi)$ with $\boldsymbol{\Sigma} > \mathbf{0}$ and $\mathbf{x}^{(1)}: m \times 1$, $0 < m < n$. Then $\phi_{q(\mathbf{x}^{(2)})}$ does not depend on the value of $\mathbf{x}^{(2)}$ if and only if $\mathbf{x}$ is normally distributed.

Theorem 4.13
Assume $\mathbf{x} = \begin{pmatrix} \mathbf{x}^{(1)} \\ \mathbf{x}^{(2)} \end{pmatrix} \sim EC_n(\boldsymbol{\mu}, \boldsymbol{\Sigma}, \phi)$ with $\boldsymbol{\Sigma} > \mathbf{0}$ and $\mathbf{x}^{(1)}: m \times 1$, $0 < m < n$. Then $(\mathbf{x}^{(1)} | \mathbf{x}^{(2)})$ is normally distributed with probability one if and only if $\mathbf{x}$ is normally distributed.

Theorem 4.14
A distribution $EC_n(\boldsymbol{\mu}, \boldsymbol{\Sigma}, \phi)$ with rank $(\boldsymbol{\Sigma}) \geqslant 2$ is a normal distribution if and only if two marginal densities of different dimensions exist and have functional forms which agree up to a positive multiple.

Theorem 4.15

Suppose $\mathbf{x} = \begin{pmatrix} \mathbf{x}^{(1)} \\ \mathbf{x}^{(2)} \end{pmatrix} \sim EC_n(\boldsymbol{\mu}, \boldsymbol{\Sigma}, \phi)$ with $\boldsymbol{\Sigma} > \mathbf{0}$ and $\mathbf{x}^{(1)}: m \times 1$, $0 < m < n$. Then, for any fixed positive integer p, $E((R_{q(\mathbf{x}^{(2)})} p | \mathbf{x}^{(2)}))$ is finite and degenerate (i.e., independent of $\mathbf{x}^{(2)}$) if and only if $\mathbf{x}$ is normally distributed.

Theorem 4.16

Suppose $\mathbf{x} = \begin{pmatrix} \mathbf{x}^{(1)} \\ \mathbf{x}^{(2)} \end{pmatrix} \sim EC_n(\boldsymbol{\mu}, \boldsymbol{\Sigma}, \phi)$ with $\boldsymbol{\Sigma} > \mathbf{0}$, $\mathbf{x}^{(1)}: m \times 1$, $0 < m < n$, and p is a positive integer. Then a pth-order conditional central moment of $\mathbf{x}^{(1)}$ given $\mathbf{x}^{(2)}$ is finite and degenerate (i.e., independent of $\mathbf{x}^{(2)}$) if and only if $\mathbf{x}$ is normally distributed.

The last five theorems (4.12–4.16) are due to Cambanis, Huang and Simons (1981). Proofs of these theorems can be found in their paper.

Another aspect of characterizations of normality is based on independence of two quadratic forms.

Lemma 4.1

Let $\mathbf{x} \sim S_n(g)$ be partitioned into $\mathbf{x} = (\mathbf{x}^{(1)\prime}, \ldots, \mathbf{x}^{(k)\prime})'$, where $\mathbf{x}^{(1)}, \ldots, \mathbf{x}^{(k)}$ have respectively $n_1, \ldots, n_k$ components of $\mathbf{x}$. Then the joint density of $y_i = \mathbf{x}^{(i)\prime}\mathbf{x}^{(i)}$, $i = 1, \ldots, k$, is given by

$$\frac{\pi^{n/2}}{\prod_1^k \Gamma(n_i/2)} \prod_1^k y_i^{n_i/2 - 1} g\left(\sum_1^k y_i\right). \tag{4.17}$$

PROOF Since the density function of $\mathbf{x}^{(1)}, \ldots, \mathbf{x}^{(k)}$ is $g(\mathbf{x}'\mathbf{x}) = g(\sum_1^k \mathbf{x}^{(i)\prime}\mathbf{x}^{(i)})$, for each nonnegative measurable function $h(\cdot)$ we have

$$\begin{aligned} Eh(y_1, \ldots, y_k) &= \int h(\mathbf{x}^{(1)\prime}\mathbf{x}^{(1)}, \ldots, \mathbf{x}^{(k)\prime}\mathbf{x}^{(k)}) g\left(\sum_1^k \mathbf{x}^{(i)\prime}\mathbf{x}^{(i)}\right) \mathrm{d}\mathbf{x}^{(1)} \cdots \mathrm{d}\mathbf{x}^{(k)} \\ &= \frac{\pi^{n/2}}{\prod_1^k \Gamma(n_i/2)} \int_{R_+^k} h(y_1, \ldots, y_k) g\left(\sum_1^k y_i\right) \prod_1^m y_i^{n_i/2 - 1} \, \mathrm{d}y_i. \end{aligned}$$

The last equality is obtained by using (1.32) which implies (4.17). □

The distribution given by (4.17) is a special case of the multivariate Liouville distributions to be discussed in detail in Chapter 6. The following theorem shows that independence of the two quadratic forms of a spherical distribution implies normality.

Theorem 4.17

Let $\mathbf{x} = \begin{pmatrix} \mathbf{x}^{(1)} \\ \mathbf{x}^{(2)} \end{pmatrix} \sim S_n(\phi)$, where $\mathbf{x}^{(1)}: m \times 1$, $1 \leqslant m < n$. Then $\mathbf{x}^{(1)'}\mathbf{x}^{(1)}$ and $\mathbf{x}^{(2)'}\mathbf{x}^{(2)}$ are independent if and only if $\mathbf{x} \sim N_n(\mathbf{0}, \sigma^2 \mathbf{I}_n)$, for some $\sigma \geqslant 0$.

PROOF The case when $\sigma = 0$ and the 'if part' of the theorem are trivial. We shall now prove the 'only if' part. Assume that $\| \mathbf{x}^{(1)} \|^2$ and $\| \mathbf{x}^{(2)} \|^2$ are independent. Let $\mathbf{U}$ and $\mathbf{V}$ be two matrices uniformly distributed† on $O(m)$ and $O(n-m)$ with c.d.f.s $F(\mathbf{V})$ and $G(\mathbf{V})$, respectively.

The joint c.f. of $\mathbf{x}^{(1)}$ and $\mathbf{x}^{(2)}$ is $\phi(\mathbf{t}'\mathbf{t}) = \phi(\| \mathbf{t}^{(1)} \|^2 + \| \mathbf{t}^{(2)} \|^2)$, where $\mathbf{t} = \begin{bmatrix} \mathbf{t}^{(1)} \\ \mathbf{t}^{(2)} \end{bmatrix} : n \times 1$ and $\mathbf{t}^{(1)}: m \times 1$. Since $\mathbf{U}$ and $\mathbf{V}$ are uniformly distributed on $O(m)$ and $O(n-m)$, respectively, we have $\mathbf{U}'\mathbf{U} = \mathbf{I}_m$ and $\mathbf{V}'\mathbf{V} = \mathbf{I}_{n-m}$. Thus

$$
\begin{aligned}
\phi(\mathbf{t}'\mathbf{t}) &= \int_{O(m)} \int_{O(n-m)} \phi(\| \mathbf{V}\mathbf{t}^{(1)} \|^2 + \| \mathbf{V}\mathbf{t}^{(2)} \|^2)\, \mathrm{d}F(\mathbf{U})\, \mathrm{d}G(\mathbf{V}) \\
&= E\Bigg[\int_{O(m)} \int_{O(n-m)} \exp\{ i\mathbf{t}^{(1)'}\mathbf{U}'\mathbf{x}^{(1)} \\
&\qquad + i\mathbf{t}^{(2)'}\mathbf{V}'\mathbf{x}^{(2)} \}\, \mathrm{d}F(\mathbf{U})\, \mathrm{d}G(\mathbf{V}) \Bigg] \\
&= E\Bigg[\int_{O(m)} \exp\{ i\mathbf{t}^{(1)'}\mathbf{U}'\mathbf{x}^{(1)} \}\, \mathrm{d}F(\mathbf{U}) \\
&\qquad \times \int_{O(n-m)} \exp\{ i\mathbf{t}^{(2)'}\mathbf{V}'\mathbf{x}^{(2)} \}\, \mathrm{d}G(\mathbf{V}) \Bigg] \\
&= E\Bigg[\int_{O(m)} \exp\{ itr(\mathbf{t}^{(1)}\mathbf{x}^{(1)'}\mathbf{U})\, \mathrm{d}F(\mathbf{U}) \} \\
&\qquad \times \int_{O(n-m)} \exp\{ itr(\mathbf{t}^{(2)}\mathbf{x}^{(2)'}\mathbf{V})\, \mathrm{d}G(\mathbf{V}) \} \Bigg]
\end{aligned}
$$

† A random matrix $\mathbf{X}: n \times n$ is said to have the uniform distribution on $O(n)$ if $\mathbf{X}'\mathbf{X} = \mathbf{I}_n$ and $\boldsymbol{\Gamma}\mathbf{X} \overset{d}{=} \mathbf{X}$ for each $\boldsymbol{\Gamma} \in O(n)$. By an argument similar to that used for the proof of Theorem 2.1, the c.f. of $\mathbf{X}$ must be of form $\Omega_n(\mathbf{T}'\mathbf{T})$, where $\mathbf{T}: n \times n$. See Fang and Zhang (1989) for more detail.

By the definition of the c.f. of random matrix, the last expectation becomes

$$E[\Omega_m(\mathbf{t}^{(1)}\mathbf{x}^{(1)'}\mathbf{x}^{(1)'}\mathbf{t}^{(1)'})\Omega_{n-m}(\mathbf{t}^{(2)}\mathbf{x}^{(2)'}\mathbf{x}^{(2)}\mathbf{t}^{(2)'})].$$

Here Ω_m and Ω_{n-m} are the c.f.s of $\mathbf{U}$ and $\mathbf{V}$, respectively. The independence of $\|\mathbf{x}^{(1)}\|^2$ and $\|\mathbf{x}^{(2)}\|^2$ implies

$$\begin{aligned}\phi(\|\mathbf{t}^{(1)}\|^2 + \|\mathbf{t}^{(2)}\|^2) \\ = E[\Omega_m(\|\mathbf{x}^{(1)}\|^2\mathbf{t}^{(1)}\mathbf{t}^{(1)'})]E[\Omega_{n-m}(\|\mathbf{x}^{(2)}\|^2\mathbf{t}^{(2)}\mathbf{t}^{(2)'})].\end{aligned}$$

By the same argument the right-hand side becomes

$$\phi(\|\mathbf{t}^{(1)}\|^2)\phi(\|\mathbf{t}^{(2)}\|^2),$$

which implies

$$\phi(\|\mathbf{t}^{(1)}\|^2 + \|\mathbf{t}^{(2)}\|^2) = \phi(\|\mathbf{t}^{(1)}\|^2)\phi(\|\mathbf{t}^{(2)}\|^2).$$

By the proof of Theorem 4.11, $\phi(u) = e^{-\sigma^2 u}$ for some $\sigma^2 > 0$, which completes the proof. □

The following corollary is of special historical and practical interest and value. A version of this theorem is a classical result of Fisher (1925).

Corollary
Let $\mathbf{x} = (x_1, \ldots, x_n)' \sim S_n(\phi)$ be nonsingular and

$$\bar{x} = \frac{1}{n}\sum_1^n x_i, \quad s^2 = \sum_1^n (x_i - \bar{x})^2.$$

Then $\bar{x}$ and s^2 are independent if and only if $\mathbf{x}$ is normal $N_n(\mathbf{0}, \sigma^2\mathbf{I}_n)$ for some $\sigma > 0$.

PROOF Let $\mathbf{\Gamma} \in O(n)$ with the last row being $(1/n^{1/2}, \ldots, 1/n^{1/2})$ and $\mathbf{y} = (y_1, \ldots, y_n)' = \mathbf{\Gamma}\mathbf{x}$. Then $\mathbf{y} \sim S_n(\phi)$ and

$$\bar{x} = n^{-1/2} y_n \quad \text{and} \quad s^2 = \sum_1^{n-1} y_i^2.$$

The independence of $\bar{x}$ and s^2 implies the independence of $\sum_1^{n-1} y_i^2$ and y_n^2, which in turn implies that $\mathbf{y}$ is normal by Theorem 4.17, i.e., $\mathbf{x} \overset{d}{=} \mathbf{y}$ is normal. □

Ali (1980) and Arnold and Lynch (1982) proved this corollary in different ways. Actually, this is an application of Theorem 4.17, which is due to Anderson and Fang (1987).

For $\mathbf{x} \sim S_n^+(\phi)$, Anderson and Fang (1982) derive the distributions of quadratic form $\mathbf{x}'\mathbf{A}\mathbf{x}$ and obtain Cochran's theorem in this general case. Khatri and Mukerjee (1987) gave a slight extension without the condition $P(\mathbf{x}=\mathbf{0})=0$. Here we state this theorem without a proof.

Theorem 4.18

Let $\mathbf{x} \sim S_n(\phi)$ and $\mathbf{A}$ be an $n \times n$ symmetric matrix. Then $\mathbf{x}'\mathbf{A}\mathbf{x} \sim x_k^2$ if and only if $\mathbf{x} \sim N_n(\mathbf{0}, \sigma^2\mathbf{I}_p)$ for some $\sigma^2 > 0$, $\mathbf{A}^2 = \sigma^{-2}\mathbf{A}$ and rank $(\mathbf{A}) = k$.

Problems

4.1 Let $\mathbf{x}$ be an m-dimensional random vector. Then $\mathbf{x} \in D(m, n)$ (cf. Definition 4.3) if and only if $\mathbf{x}$ has a density $h(\|\mathbf{x}\|^2)$, where

$$h(t) = \int_0^\infty g(t/r; m, n) r^{-m/2} \, \mathrm{d}G(r),$$

where

$$\mathrm{g}(t; m, n) = C_{m,n}(1-t)^{n/2-1},$$

$$C_{m,n} = \Gamma\left(\frac{m+n}{2}\right) \Big/ \left(\Gamma\left(\frac{n}{2}\right)\pi^{m/2}\right),$$

and G is a c.d.f. with $G(0)=0$ (cf. Eaton, 1981).

4.2 Suppose $\mathbf{y} = (y_1, \ldots, y_n)'$ is a vector of independent components and that a $\boldsymbol{\Gamma}$ is a random orthogonal matrix with the distribution having the support $\mathcal{O}(n)$. Then the y_i are identically distributed $N(0, \sigma^2)$ for some σ^2 if and only if (i) $\boldsymbol{\Gamma}\mathbf{y}$ is independent of $\boldsymbol{\Gamma}$, or (ii) $\boldsymbol{\Gamma}\mathbf{y} \overset{d}{=} \mathbf{y}$ (cf. Brown, Cartwright and Eagleson, 1985).

4.3 Provide detailed proofs of Theorems 4.14–4.16.

4.4 A distribution function $F(\mathbf{x})$ defined on R^n is said to be α-unimodal with a mode of $\mathbf{0}$ if the function

$$S(t; \alpha, g, F) = t^\alpha \int_{R^n} g(t\mathbf{x}) \, \mathrm{d}F(\mathbf{x})$$

is nondecreasing in t for $t > 0$ for every bounded, nonnegative Borel measurable function g on R^n.

Show that

(i) every n-dimensional spherically symmetric α-unimodal distribution function with a mode at $\mathbf{0}$ is absolutely continuous on $R^n - \{\mathbf{0}\}$;

(ii) the convolution of two n-dimensional spherical, α-unimodal distributions is α-unimodal.

(Cf. Wolfe, 1975.)

4.5 Assume that $\mathbf{x}$ is an $n \times 1$ random vector for which there are constant nonzero vectors $\boldsymbol{\alpha}$ and $\boldsymbol{\beta}$ such that $\boldsymbol{\alpha}'\boldsymbol{\Gamma}\mathbf{x}$ and $\boldsymbol{\beta}'\boldsymbol{\Gamma}\mathbf{x}$ are independent for each $\boldsymbol{\Gamma} \in \mathcal{O}(n)$. Then either the $\{x_i\}$ are degenerate or $\boldsymbol{\alpha}'\boldsymbol{\beta} = 0$ and the components of $\mathbf{x}$ are independent normal variables with a common variance (cf. Berk, 1986).

4.6 Assume that $\mathbf{x} \sim S_n(\phi)$ and there exist two independent linear combinations $\boldsymbol{\alpha}'\mathbf{x}$ and $\boldsymbol{\beta}\mathbf{x}(\boldsymbol{\alpha} \neq \mathbf{0} \neq \boldsymbol{\beta})$. Then $\mathbf{x} \sim N_n(\mathbf{0}, \sigma^2\mathbf{I}_n)$ for some $\sigma^2 > 0$. This result is related to the well-known Darmois–Skitovich theorem for the normal distributions.

CHAPTER 5

Multivariate ℓ_1-norm symmetric distributions

In this chapter we shall introduce several families of multivariate ℓ_1-norm symmetric distributions, which include the i.i.d. sample from exponential as its particular case. One family, denoted by n, consists of the scale mixture of the uniform distributions on the surface of the ℓ_1-norm unit sphere. Some of its properties will be discussed. Also described are a more general family T_n, of which the survival functions are functions of ℓ_1-norm, and an important subset of L_n, namely $L_{n,\infty}$, which is constructed as a scale mixture of random vectors with i.i.d. exponential components. The relationships among these three families and some applications are given.

The content of this chapter is based on a series of papers by K.T. Fang and B.Q. Fang (1987a, 1987b, 1988, 1989).

5.1 Definition of L_n

As can be seen from Chapter 2, the main route for development of the theory of elliptically symmetric distributions can be represented schematically as follows:

$$\begin{gathered} x_i \Rightarrow \mathbf{x} = \begin{bmatrix} x_1 \\ \vdots \\ x_n \end{bmatrix} \Rightarrow \mathbf{u}^{(n)} = \mathbf{x}/\|\mathbf{x}\| \Rightarrow r\mathbf{u}^{(n)} \Rightarrow \boldsymbol{\mu} + r\mathbf{A}'\mathbf{u}^{(n)} \\ N(0,1) \Rightarrow N_n(\mathbf{0}, \mathbf{I}_n) \Rightarrow \text{uniform} \Rightarrow \text{spherical} \Rightarrow \text{elliptical}. \end{gathered} \tag{5.1}$$

Using this approach, we were able to generalize properties of the multinormal distribution to spherically and elliptically symmetric distributions.

Using a similar approach, we shall develop several families of ℓ_1-norm symmetric distributions related to the exponential distribution, which are important and widely used distributions in statistics. We should remark that there are other kinds of multivariate exponential distributions which have been defined and studied in the literature: Gumbel (1960), Block (1985), Marshall and Olkin (1967a, 1967b), Raftery (1984), and Lindley and Singpurwalla (1986). For a comprehensive survey see, e.g. Johnson and Kotz (1972), Block (1985) and Marshall and Olkin (1985).

Motivated by (5.1) we present and develop a new type of a multivariate extension of the exponential distribution. This family is constructed as follows. Let $x_1,\ldots,x_n$ be an i.i.d. sample from an exponential distribution $E(\lambda)$, i.e., the density of x_1 is given by

$$\begin{cases} \lambda e^{-\lambda x_1}, & x_1 > 0 \\ 0, & \text{otherwise.} \end{cases} \tag{5.2}$$

Let $\mathbf{x} = (x_1,\ldots,x_n)'$. The ℓ_1-norm of $\mathbf{x}$ is $\|\mathbf{x}\|_1 = x_1 + \cdots + x_n$. Hence in the spirit of relation (5.1) (without the last step):

$$\begin{array}{c} x_i \Rightarrow \mathbf{x} = \begin{bmatrix} x_1 \\ \vdots \\ x_n \end{bmatrix} \Rightarrow \mathbf{u} = \mathbf{x}/\|\mathbf{x}\|_1 \Rightarrow r\mathbf{u} \\ E(1) \Rightarrow \text{i.i.d. } E(1) \Rightarrow \text{uniform} \Rightarrow \ell_1\text{-norm symmetric.} \end{array} \tag{5.3}$$

We shall proceed in this chapter along the lines of (5.3). Throughout this chapter, let $x_1,\ldots,x_n$ be i.i.d., $x_1 \sim E(1)$ and let $R^n_+ = \{(z_1,\ldots,z_n)' : z_i \geqslant 0,\ i = 1,\ldots,n\}$ as before, and

$$B_n = \{\mathbf{z} \mid \mathbf{z} \in R^n_+,\ \|\mathbf{z}\|_1 = 1\}. \tag{5.4}$$

To simplify the notation, we shall use $\|\cdot\|$ in place of $\|\cdot\|_1$ in this chapter. Let

$$\mathbf{u} = \begin{bmatrix} u_1 \\ \vdots \\ u_n \end{bmatrix} \overset{d}{=} \mathbf{x}/\|\mathbf{x}\|. \tag{5.5}$$

Since $E(1)$ is a particular case of the gamma distribution, i.e., $E(1) = \text{Ga}(1)$ (cf. (1.16)), $\mathbf{u} \sim D_n(1,\ldots,1)$ (cf. Definition 1.4), where $\sum_{i=1}^n u_i = 1$ and the density of $u_1,\ldots,u_{n-1}$ is given by

$$p_{n-1}(u_1,\ldots,u_{n-1}) = \begin{cases} (n-1)! & \text{for } (u_1,\ldots,u_{n-1})' \in A_{n-1}, \\ 0 & \text{otherwise} \end{cases} \tag{5.6}$$

where $A_n = \{\mathbf{x} : \mathbf{x} \in R^n_+,\ \|\mathbf{x}\| \leqslant 1\}$. Thus $\mathbf{u}$ is uniformly distributed on

B_n, the surface of the ℓ_1-norm unit sphere, and write $\mathbf{u} \sim \nu_n$. Let

$$L_n = \{\mathbf{z} | \mathbf{z} \stackrel{d}{=} r\mathbf{u},\ r \geqslant 0 \text{ and is independent of } \mathbf{u} \sim \nu_n\}. \quad (5.7)$$

Definition 5.1
Let $\mathbf{z}$ be an n-dimensional random vector. If $\mathbf{z} \in L_n$, we say that $\mathbf{z}$ has a **multivariate ℓ_1-norm symmetric** distribution and call $\mathbf{u}$ the **uniform base** and r the **generating variate.** The distribution function $F(\cdot)$ and the density $f(\cdot)$ (if it exists) of r are respectively called the **generating c.d.f.** and **generating density** of $\mathbf{z}$.

5.2 Some properties of L_n

Several basic properties of L_n, pertaining to survival functions, characteristic functions, probability density functions, marginal distributions, conditional distributions and moments, etc., are presented in this section.

5.2.1 *Survival function and characteristic function*

Theorem 5.1
If $\mathbf{z} = r\mathbf{u} \in L_n$, $P(\mathbf{z} = \mathbf{0}) = 0$, then

$$\|\mathbf{z}\| \stackrel{d}{=} r, \qquad \mathbf{z}/\|\mathbf{z}\| \stackrel{d}{=} \mathbf{u},$$

and they are independent.

The proof of Theorem 5.1 is very similar to that of Theorem 2.3 and it is therefore omitted, leaving the reader to fill in the details.

The lemma below deals with several integration formulas. Here we also omit proofs since they can easily be derived using the results of Section 1.4.

Lemma 5.1

$$\int_{A_n} \cdots \int x_1^{p_1 - 1} \ldots x_n^{p_n - 1} \mathrm{d}x_1 \ldots \mathrm{d}x_n = \prod_{k=1}^n \Gamma(p_k) / \Gamma\left(\sum_{k=1}^n p_k + 1\right), \quad (5.8)$$

where $p_k > 0$, $k = 1, \ldots, n$, and A_n is as defined in the last section;

$$\int_{A_n(b)} \cdots \int \mathrm{d}x^1 \ldots \mathrm{d}x_n = (\Gamma(n+1))^{-1} b^n, \quad (5.9)$$

where $b > 0$ and $A_n(b) = \{\mathbf{x}: \mathbf{x} \in R^n_+, \|\mathbf{x}\| \leqslant b\}$ (obviously, $A_n = A_n(1)$);

$$\int_{R^n_+} \cdots \int f(\|\mathbf{z}\|)\,\mathrm{d}z_1 \ldots \mathrm{d}z_n = (\Gamma(n))^{-1} \int_0^\infty f(x) x^{n-1}\,\mathrm{d}x, \qquad (5.10)$$

where $f(\|\mathbf{z}\|)$ is an integrable function on R^n_+.

The definition of L_n suggests that we should perhaps focus our attention in the first place on the uniform distribution v_n.

Theorem 5.2
Suppose $\mathbf{u} \sim v_n$. Then

(1) $P(u_1 > a_1, \ldots, u_n > a_n) = (1 - \|\mathbf{a}\|)^{n-1} I_{A_n}(\mathbf{a}) \equiv h_0(\|\mathbf{a}\|)$,

where $I_{A_n}(\cdot)$ is the indicator function of A_n.

(2) $(u_1, \ldots, u_m)$ has a density function

$$\frac{\Gamma(n)}{\Gamma(n-m)}\left(1 - \sum_{k=1}^m u_k\right)^{n-m-1} I_{A_n}(u_1, \ldots, u_m),$$

where $1 \leqslant m \leqslant n-1$.

This well-known result was originally derived by B. de Finetti (cf. e.g., Feller, 1971, p. 42).

PROOF Let

$$D_1 = \left\{\mathbf{y}: \mathbf{y} \in R^n_+, y_k > a_k,\ k = 1, \ldots, n-1,\ \sum_{k=1}^{n-1} y_k < 1 - a_n\right\}$$

and

$$D_2 = \left\{\mathbf{y}: \mathbf{y} \in R^n_+, y_k > 0,\ k = 1, \ldots, n-1,\ \sum_{k=1}^{n-1} y_k < 1 - \sum_{k=1}^{n} a_k\right\}.$$

For any $\mathbf{a} \in A_n$, we have

$$\begin{aligned} &P\{u_1 > a_1, \ldots, u_n > a_n\} \\ &= P\left\{u_1 > a_1, \ldots, u_{n-1} > a_{n-1}, \sum_{k=1}^{n-1} u_k < 1 - a_n\right\} \\ &= \Gamma(n) \int_{D_1} \cdots \int I_{A_{n-1}}(u_1, \ldots, u_{n-1})\,\mathrm{d}u_1 \ldots \mathrm{d}u_{n-1}. \end{aligned}$$

Applying the integral transformation $x_k = u_k - a_k$, $k = 1, \ldots, n-1$, and Lemma 5.1 we obtain

$$P(u_1 > a_1, \ldots, u_n > a_n)$$
$$= \Gamma(n) \int_{D_2} \cdots \int \mathrm{d}x_1 \ldots \mathrm{d}x_{n-1} = (1 - \|\mathbf{a}\|)^{n-1}.$$

The second assertion follows directly from Theorems 1.5 and 1.2. □

From Theorem 5.2(2), we deduce the following results:

1. $(u_1, \ldots, u_m) \sim D_{m+1}(1, \ldots, 1, n-m)$, $1 \leqslant m \leqslant n-1$
2. $u_k \sim \mathrm{Be}(1, n-1)$, $k = 1, \ldots, m$.

Theorem 5.3
Let $\mathbf{u} \sim v_n$. Then the c.f. of $\mathbf{u}$ is given by

$$\phi_0(t_1, \ldots, t_n) = e^{it_n} \sum_{m=0}^{\infty} \frac{i^m}{(n+m-1)^{(m)}} \sum_{p_1 + \cdots + p_{n-1} = m} \prod_{k=1}^{n-1} (t_k - t_n)^{p_k}. \tag{5.11}$$

PROOF Lebesgue's dominated convergence theorem yields

$$E(e^{i\mathbf{t}'\mathbf{u}}) = E\left[\exp\left[i \sum_{k=1}^{n-1} t_k u_k + it_n\left(1 - \sum_{k=1}^{n-1} u_k\right)\right]\right]$$
$$= e^{it_n} \sum_{m=0}^{\infty} \frac{i^m}{m!} \sum_{p_1 + \cdots + p_{n-1} = m} \frac{m!}{p_1! \ldots p_{n-1}!} \prod_{k=1}^{n-1} (t_k - t_n)^{p_k}$$
$$\times \int_{A_{n-1}} \cdots \int \Gamma(n) u_1^{p_1} \ldots u_{n-1}^{p_{n-1}} \mathrm{d}u_1 \ldots \mathrm{d}u_{n-1}.$$

Hence the assertion follows from Lemma 5.1. □

Theorem 5.4
If $\mathbf{z} \stackrel{d}{=} r\mathbf{u} \in L_n(F)$, $F(x)$ being the generating c.d.f. of $\mathbf{z}$, then

$$P(z_1 > a_1, \ldots, z_n > a_n) = \int_{\|\mathbf{a}\|}^{\infty} (1 - \|\mathbf{a}\|/r)^{n-1} \mathrm{d}F(r) \tag{5.12}$$

where $\mathbf{a} \in R_+^n$, and

$$E(e^{i\mathbf{t}'\mathbf{z}}) = \int_0^{\infty} \phi_0(rt_1, \ldots, rt_n) \mathrm{d}F(r), \tag{5.13}$$

where the c.f. $\phi_0(\cdot)$ is given by (5.11).

PROOF Formulas (5.12) and (5.13) follow directly from Theorem 5.2. □

Corollary
The distribution function of $\mathbf{z} \in L_n$ is uniquely determined by its one-dimensional marginal distribution.

PROOF Indeed

$$P(z_1 > u_1, \ldots, z_n > u_n) = P\left(z_j > \sum_{i=1}^{n} u_i\right), \qquad j = 1, \ldots, n, \tag{5.14}$$

$(u_1, \ldots, u_n)' \in R_+^n$. □

Let $\mathbb{R}$ be the set of all nonnegative random variables, i.e., $\mathbb{R} = \{r: r$ is a nonnegative random variable$\}$. Let T be a map from $\mathbb{R}$ to L_n with $T(r) = r\mathbf{u}$, where $r \in \mathbb{R}$ and $\mathbf{u} \sim v_n$ are independent. Since $T(r) = T(q)$ for r and $q \in \mathbb{R}$ implies $r = \| r\mathbf{u} \| \stackrel{d}{=} \| q\mathbf{u} \| = q$, the mapping T is injective. Moreover, T is also surjective since for any $\mathbf{z} \in L_n$ with the stochastic representation $\mathbf{z} \stackrel{d}{=} r\mathbf{u}$, we have $T(r) = \mathbf{z}$. Therefore T is a one-to-one correspondence between $\mathbb{R}$ and L_n. From now on, we shall write $\mathbf{Z} \stackrel{d}{=} r\mathbf{u} \in L_n(F)$ to mean that $\mathbf{z} \in L_n$, $\mathbf{z}$ having the stochastic representaion $\mathbf{z} \stackrel{d}{=} r\mathbf{u}$, where $r \geqslant 0$ has the distribution function $F(\cdot)$ and is independent of $\mathbf{u} \sim v_n$.

Remark: By comparing Theorem 2.12 with the above statement, the reader can devise different approaches for solving the same problems.

5.2.2 *Density*

If $\mathbf{z} \in L_n(F)$, it is not necessary for $\mathbf{z}$ to possess a density. From the discussion above, it is easy to see that the density of $\mathbf{z}$, if it exists, depends upon the density of r and conversely.

Theorem 5.5
Suppose $\mathbf{z} \stackrel{d}{=} r\mathbf{u} \in L_n(F)$. Then $\mathbf{z}$ possesses a density on $R_+^n - \{0\}$ iff r has a density on $(0, \infty)$. In this case, the density of $\mathbf{z}$ is of the form $g(\sum_{k=1}^n z_k)$ on $R_+^n - \{0\}$, where $g(\cdot)$ is related to the density of F in the following way:

$$g(x) = \Gamma(n) x^{-n+1} f(x), \qquad x > 0, \tag{5.15}$$

where $f(x) = F'(x)$.

PROOF Assume r possesses a p.d.f. $f(r)$ on $(0, \infty)$. The joint density of $r, u_1, \ldots, u_{n-1}$ is given by

$$f(r)\cdot\Gamma(n)I_{A_{n-1}}(u_1, \ldots, u_{n-1}).$$

Using the transformation

$$\begin{cases} z_k = ru_k, & k = 1, \ldots, n-1 \\ z_n = r(1 - u_1 - \cdots - u_{n-1}), \end{cases}$$

we arrive at the p.d.f. of $\mathbf{z}$ of the form:

$$g(\|\mathbf{z}\|) = \Gamma(n)\|\mathbf{z}\|^{-n+1}f(\|\mathbf{z}\|) \quad \text{for} \quad \mathbf{z}\in R^n_+ - \{0\},$$

i.e., $\mathbf{z}$ possesses a density on $R^n_+ - \{0\}$ and $f(\cdot)$ and $g(\cdot)$ are given by the relationship (5.15).

Conversely, if $\mathbf{z}$ possesses a density on $R^n_+ - \{0\}$ which has the form of $g(\|\mathbf{z}\|)$, since $r \overset{d}{=} \|\mathbf{z}\|$, it follows that r also possesses a density on $(0, \infty)$. □

5.2.3 *Marginal distributions*

In this subsection, it will be shown that if $\mathbf{z}\in L_n(F)$, then its m-dimensional marginal distribution $(1 \leqslant m < n)$ belongs to L_m and possesses a density on $R^m_+ - \{0\}$, while $\mathbf{z}$ may be not absolutely continuous on $R^n_+ - \{0\}$.

Theorem 5.6
Denote $\mathbf{u} = (\mathbf{u}'_{(1)}, \ldots, \mathbf{u}'_{(k)})' \sim v_n$, where $\mathbf{u}_{(j)}$ is n_j-dimensional, $j = 1, \ldots, k$, $\sum_{j=1}^k n_j = n$. Then

$$(\mathbf{u}_{(1)}, \ldots, \mathbf{u}_{(k)}) = (w_1\mathbf{u}_1, \ldots, w_k\mathbf{u}_k),$$

where $\mathbf{w} = (w_1, \ldots, w_k)'$, $\mathbf{u}_1, \ldots, \mathbf{u}_k$ are independent, $\mathbf{u}_j \sim v_{n_j}, j = 1, \ldots, k$, and $\mathbf{w} \sim D_k(n_1, \ldots, n_k)$.

PROOF Assume $x_1, \ldots, x_n$ are i.i.d. and $\mathbf{x}_1 \sim E(1)$. Partition $\mathbf{x} = (x_1, \ldots, x_n)'$ into $\mathbf{x} = (\mathbf{x}^{(1)\prime}, \ldots, \mathbf{x}^{(k)\prime})'$ where $\mathbf{x}^{(j)}: n_j \times 1, j = 1, \ldots, k$. Then

$$\mathbf{u} \overset{d}{=} \mathbf{x}/\|\mathbf{x}\| = \begin{pmatrix} \dfrac{\|\mathbf{x}^{(1)}\|}{\|\mathbf{x}\|}\cdot\dfrac{\mathbf{x}^{(1)}}{\|\mathbf{x}^{(1)}\|} \\ \vdots \\ \dfrac{\|\mathbf{x}^{(k)}\|}{\|\mathbf{x}\|}\cdot\dfrac{\mathbf{x}^{(k)}}{\|\mathbf{x}^{(k)}\|} \end{pmatrix}.$$

Set $w_j = \|\mathbf{x}^{(j)}\|/\|\mathbf{x}\|$, $\mathbf{u}_{(j)} = \mathbf{x}^{(j)}/\|\mathbf{x}^{(j)}\|$, $j = 1,\ldots,k$. It is easy to verify that $\{w_j, \mathbf{u}_{(j)}, j = 1,\ldots,k\}$ satisfy the conditions of the theorem. □

5.2.4 *Conditional distributions*

In this subsection it will be shown that the conditional distributions of L_n are also ℓ_1-norm symmetric distributions.

Theorem 5.7
Let $\mathbf{z} \overset{d}{=} r\mathbf{u} \in L_n(F)$ and $\mathbf{z} = (\mathbf{z}^{(1)\prime}, \mathbf{z}^{(2)\prime})'$, where $\mathbf{z}^{(1)}$ is m-dimensional, $1 \leqslant m < n$. Then the regular conditional distribution of $\mathbf{z}^{(1)}$ given $\mathbf{z}^{(2)}$ is given by

$$(\mathbf{z}^{(1)} | \mathbf{z}^{(2)} = \mathbf{w}) \overset{d}{=} r_{\|\mathbf{w}\|} \mathbf{u}_1 \in L_m(F_{\|\mathbf{w}\|}),$$

where $r_{\|\mathbf{w}\|}$ and $\mathbf{u}_1$ are independent, $\mathbf{u}_1 \sim \nu_m$ and $r_{\|\mathbf{w}\|} = 0$ with probability 1 if $\|\mathbf{w}\| = 0$ or $F(\|\mathbf{w}\|) = 1$, and moreover

$$\begin{aligned} F_{\|\mathbf{w}\|}(t) &= P(r_{\|\mathbf{w}\|} \leqslant t) \\ &= \frac{\displaystyle\int_{\|\mathbf{w}\|}^{t+\|\mathbf{w}\|} (s - \|\mathbf{w}\|)^{m-1} s^{-n+1} \, \mathrm{d}F(s)}{\displaystyle\int_{\|\mathbf{w}\|}^{\infty} (s - \|\mathbf{w}\|)^{m-1} s^{-n+1} \, \mathrm{d}F(s)}, \end{aligned}$$

where $t \geqslant 0$, $\|\mathbf{w}\| > 0$, $F(\|\mathbf{w}\|) < 1$.

PROOF Theorem 5.6 yields

$$\mathbf{z} \overset{d}{=} r\mathbf{u} = r(w_1 \mathbf{u}_1, w_2 \mathbf{u}_2),$$

where $(w_1, w_2) \sim D_2(m, n-m)$, $\mathbf{u}_1 \sim \nu_m$, $\mathbf{u}_2 \sim \nu_{n-m}$, (w_1, w_2), r, $\mathbf{u}_1$ and $\mathbf{u}_2$ are independent. Then

$$\begin{aligned} (rw_1 | rw_2\mathbf{u}_2 = \mathbf{w}) &\overset{d}{=} (r - \|\mathbf{w}\| \,|\, rw_2\mathbf{u}_2 = \mathbf{w}) \\ &\overset{d}{=} (r - \|\mathbf{w}\| \,|\, rw_2 = \|\mathbf{w}\|), \end{aligned}$$

in view of the independence of r, w_2 and $\mathbf{u}_2$. Let

$$r_{\|\mathbf{w}\|} = (r - \|\mathbf{w}\| \,|\, \mathbf{z}^{(2)} = \mathbf{w}).$$

Then

$$(\mathbf{z}^{(1)}|\mathbf{z}^{(2)}=\mathbf{w})\overset{d}{=}r_{\|\mathbf{w}\|}\mathbf{u}_1\in L_m(F_{\|\mathbf{w}\|}).$$

It is easy to verify that

$$F_{\|\mathbf{w}\|}(r)=0 \quad \text{for } r<0,$$

and

$$F_{\|\mathbf{w}\|}(r)=1 \quad \text{for } \|\mathbf{w}\|=0 \quad \text{or} \quad \|\mathbf{w}\|>0,\ F(\|\mathbf{w}\|)=1.$$

Since w_2 possesses a density, say $g(\cdot)$, which is given by

$$g(x)=\frac{\Gamma(n)}{\Gamma(m)\Gamma(n-m)}x^{n-m-1}(1-x)^{m-1}I_{[0,1]}(x),$$

rw_2 also possesses a density, namely

$$f_{n-m}(x)=\int_x^\infty \frac{\Gamma(n)}{\Gamma(m)\Gamma(n-m)}(1-x/r)^{m-1}(x/r)^{n-m-1}r^{-1}\,\mathrm{d}F(r),$$

$$0<x<\infty.$$

For $\|\mathbf{w}\|>0$, $F(\|\mathbf{w}\|)<1$, $t\geqslant 0$ we have

$$F_{\|\mathbf{w}\|}(t)=\int_{\|\mathbf{w}\|}^{t+\|\mathbf{w}\|} g(\|\mathbf{w}\|/s)s^{-1}\,\mathrm{d}F(s)/f_{n-m}(\|\mathbf{w}\|),$$

which, together with the expressions for g and f_{n-m}, establishes the assertion. □

5.2.5 *Moments*

Theorem 5.8

The moments of $\mathbf{z}\overset{d}{=}r\mathbf{u}$, provided they exist, can be expressed in terms of a one-dimensional integral, i.e., we have

$$E\left(\prod_{i=1}^n z_i^{m_i}\right)=\frac{E(r^m)}{n^{[m]}}\prod_{i=1}^n (m_i)! \tag{5.16}$$

where $m_1,\ldots,m_n$ are any positive integers, $m=m_1+\cdots+m_n$.

PROOF Since $\mathbf{z}\overset{d}{=}r\mathbf{u}$, r and $\mathbf{u}$ are independent, and $\mathbf{u}\sim D_n(1,\ldots,1)$, hence,

$$E\left(\prod_{i=1}^{n} z_i^{m_i}\right) = E(r^m)E\left(\prod_{i=1}^{n} u_i^{m_i}\right).$$

The formula (5.16) now follows from Theorem 1.3. □

5.2.6 *Some examples*

Example 5.1

Let $\mathbf{z}$ have a uniform distribution in A_n, having density $g(\|\mathbf{z}\|) = \Gamma(n+1)I_{A_n}(z_1, \ldots, z_n)$. Then $\mathbf{z} \in L_n(F)$, where $\|\mathbf{z}\|$ has a density

$$f(r) = F'(r) = (\Gamma(n))^{-1} r^{n-1} g(r) = nr^{n-1} I_{[0,1]}(r).$$

Example 5.2

Suppose a nonnegative random variable r has a density $f(r)$. Let $\mathbf{z} \overset{d}{=} r\mathbf{u}$, where r and $\mathbf{u}$ are independent. Then $\mathbf{z} \in L_n(F)$ with a density $g(\|\mathbf{z}\|)$, $\mathbf{z} \in R_+^n$, where

$$g(t) = \Gamma(n) t^{-n+1} g(t), \qquad t > 0.$$

The following are some special cases:

1. If $f(r) = cr^k I_{(0,a)}(r)$, where k is a nonnegative integer, $a > 0$, $c = (k+1)/a^{k+1}$. Then

$$g(t) = c\Gamma(n) t^{-n+k+1} I_{(0,a)}(t).$$

Two special cases are $k = 0$, implying that r is uniformly distributed on $(0, a)$ and $k = n - 1$, implying that $\mathbf{z}$ is uniformly distributed on $A_n(a)$.

2. If $r \sim \mathrm{Ga}(p, 1)$, then

$$g(t) = \frac{\Gamma(n)}{\Gamma(p)} e^{-t} t^{-n+p} I_{(0,\infty)}(t).$$

In particular, if $p = n$, then $r \sim \mathrm{Ga}(n, 1)$ and

$$g(t) = e^{-t} I_{(0,\infty)}(t),$$

namely in this case

$$\mathbf{z} \overset{d}{=} \mathbf{x}.$$

3. If $r \sim \mathrm{Be}(p, q)$, then

$$g(t) = \frac{\Gamma(n)}{B(p,q)} t^{-n+p} (1-t)^{q-1} I_{(0,1)}(t).$$

In particular, when $p = n$, we have

$$g(t) = (n+q-1)^{[n]}(1-t)q^{-1}I_{(0,1)}(t).$$

4. If $r \sim F(p, q)$, then

$$g(t) = \Gamma(n)(p/q)^{p/2}t^{-n+p/2}\left(1+\frac{p}{q}t\right)^{-(p+q)/2} \Big/ B(p/2, q/2), \quad t > 0,$$

5.3 Extended T_n family

It has been shown in the discussion above that, for $\mathbf{z} \in L_n(F)$, the probability of $(z_1 > u_1, \ldots, z_n > u_n)$ is given by the formula (5.12) which is a function of $\|\mathbf{z}\|$. Taking heed of this result, one may consider a more general family defined as:

$$T_n = \{\mathbf{z}: \mathbf{z} \in R_+^n, P(z_1 > u_1, \ldots, z_n > u_n) = h(\|\mathbf{z}\|)\},$$

where $h(\cdot)$ is a function on $[0, \infty)$.

The notation $\mathbf{z} \in T_n(h)$ will be used if we wish to emphasize the function $h(\cdot)$.

5.3.1 *Relationships between L_n and T_n families of distributions*

Clearly $L_n \subseteq T_n$. If $\mathbf{z}$ possesses a density, the following theorem provides us with a relationship between L_n and T_n.

Theorem 5.9†

Suppose $\mathbf{z} \in R_+^n$, $P(\mathbf{z} = \mathbf{0}) = 0$. Then the following statements are equivalent:

(1) $\mathbf{z} \stackrel{d}{=} r\mathbf{u} \in L_n$ and r possesses the density $f(\cdot)$.

(2) $\mathbf{z} \in T_n(h)$ and the n-dimensional function $h(\|\mathbf{w}\|)$ is absolutely continuous.

(3) $\mathbf{z}$ possesses a density of the form $g(\|\mathbf{z}\|)$, $\mathbf{z} \in R_+^n$.

In this case, g, h and f are related as follows:

$$\Gamma(n)x^{-n+1}f(x) = g(x) = (-1)^n h^{(n)}(x), \qquad x > 0 \tag{5.17}$$

$$h(t) = (\Gamma(n))^{-1}\int_0^\infty g(x+t)x^{n-1}\,\mathrm{d}x, \qquad t \geqslant 0. \tag{5.18}$$

† In this theorem and in Theorems 5.10 and 5.11 below, $\mathbf{z} \in L_n(F)$ means that $\mathbf{z} \in L_n(F)$ for some F (not a specific F). A similar remark is valid for the notation $\mathbf{z} \in T_n(h)$.

PROOF If (1) holds, then the joint density of $r, u_1, \ldots, u_{n-1}$ is given by

$$f(r)\Gamma(n)I_{A_{n-1}}(u_1, \ldots, u_{n-1}).$$

Transforming

$$\begin{cases} z_k = ru_k & k = 1, \ldots, n-1, \\ z_n = r(1 - u_1 - \cdots - u_{n-1}), \end{cases}$$

we arrive at the density of $\mathbf{z}$ given by

$$g(\|\mathbf{z}\|) = \Gamma(n)\|\mathbf{z}\|^{n+1}f(\|\mathbf{z}\|), \qquad \mathbf{z} \in R^n_+.$$

Thus (3) is valid and the relation for $f(\cdot)$ and $g(\cdot)$ holds.

Under the assumption (3), carrying out the same transformation as above we can easily establish (1). Thus (1) and (3) are equivalent.

It is also easy to show that $h^{(n)}(t)$ exists for $t \geqslant 0$ iff $(\partial^n/\partial z_1 \ldots \partial z_n)$ $h(\|\mathbf{z}\|)$ exists for $\mathbf{z} \in R^n_+$ and in this case,

$$h^{(n)}(\|\mathbf{z}\|) = \frac{\partial^n}{\partial z_1 \ldots \partial z_n} h(\|\mathbf{z}\|).$$

Now let (1) and (3) hold. Then by Theorem 5.4 $\mathbf{z} \in T_n(h)$ for some $h(\cdot)$ with the density of the form $g(\|\mathbf{z}\|)$. Let $\mathbf{w} = -\mathbf{z}$. Then $\mathbf{w}$ has the density

$$\frac{\partial^n}{\partial w_1 \ldots \partial w_n} h\left(-\sum_{k=1}^n w_k\right) = (-1)^n h^{(n)}\left(-\sum_{k=1}^n w_k\right),$$
$$w_k < 0, \ 1 \leqslant k \leqslant n.$$

Hence (2) and the second equation of (5.17) hold.

Now if (2) holds, then $\mathbf{z}$ has the density

$$g(\|\mathbf{z}\|) = (-1)^n h^{(n)}(\|\mathbf{z}\|), \qquad \mathbf{z} \in R^n_+,$$

as was shown above. This implies (3). Direct calculations yield (5.18). □

Theorem 5.9 suggests that the set L_n and T_n are equivalent provided $\mathbf{z}$ possesses a density. We shall now consider the case when the density of $\mathbf{z}$ does not exist. To investigate the relationship between L_n and T_n in this case the concept of an n-times monotone function will be required.

Definition 5.2

A function $f(t)$ on $(0, \infty)$ is called ***n*-times monotone** where n is an integer, $n \geqslant 2$, provided $(-1)^k f^{(k)}(t)$ is nonnegative, nonincreasing and convex on $(0, \infty)$, $k = 0, 1, 2, \ldots, n-2$. When $n = 1$, $f(t)$ is 1-time monotone if it is nonnegative and nonincreasing (cf. e.g. Williamson, 1956).

Theorem 5.10

$L_1 = T_1$. Moreover, if $n \geqslant 2$, then the following statements are equivalent:

(1) $\mathbf{z} \in L_n(F)$ with $F(0) = 0$.
(2) $\mathbf{z} \in T_n(h)$ with $h(0) = 1$, h being n-times monotone on $(0, \infty)$.

PROOF Clearly, $L_1 = T_1 = \mathbb{R}$. When $n \geqslant 2$, under the assumption stipulated in (1) we obtain from Theorem 5.4 that $\mathbf{z} \in T_n(h)$ with

$$h(t) = \int_t^\infty (1 - t/r)^{n-1} \, \mathrm{d}F(r), \qquad t \geqslant 0.$$

Thus h is an n-times monotone function (cf. Williamson, 1956). Moreover,

$$h(0) = P(z_1 > 0) = 1.$$

This proves assertion (2).

Conversely, if (2) holds, then

$$h(t) = \int_0^\infty (1 - st)^{n-1} I_{(0,1)}(st) \mathrm{d}v(s), \qquad t > 0,$$

where $v(s)$ is nondecreasing and $v(0) = 0$ (cf. Williamson, 1956). Noting that $1 = h(0) = h(0+) \geqslant v(+\infty)$, we obtain

$$1 = h(0+) = \int_0^\infty \mathrm{d}v(s) = v(\infty).$$

Hence

$$\mathbf{z} \stackrel{d}{=} (1/s)\mathbf{u} \in L_n,$$

where s has the d.f. $v(s)$ and $P(\mathbf{z} = \mathbf{0}) = 0$, yielding (1). □

Theorem 5.11
Let $G(\cdot)$ be the c.d.f. of a nonnegative random variable, $h(\cdot)$ be a function defined on $[0, \infty)$ and $f(\cdot)$ be the gamma density with shape parameter n. Then

$$h(t) = \int_{(t,\infty)} (1 - t/r)^{n-1} \mathrm{d}G(r), \qquad t \geqslant 0 \tag{5.19}$$

iff

$$\int_{(0,\infty)} h(s/x) f(x) \mathrm{d}x = \int_{(0,\infty)} e^{-s/r} \mathrm{d}G(r), \qquad s > 0. \tag{5.20}$$

PROOF If (5.19) holds, then

$$\begin{aligned}\int_{(0,\infty)} h(s/x) f(x) \mathrm{d}x &= \int_{(0,\infty)} \left[\int_{(0,\infty)} h_0(s/xr) \mathrm{d}G(r) \right] f(x) \mathrm{d}x \\ &= \int_{(0,\infty)} \left[\int_{(0,\infty)} h_0(s/xr) f(x) \mathrm{d}x \right] \mathrm{d}G(r) \\ &= \int_{(0,\infty)} e^{-s/r} \mathrm{d}G(r).\end{aligned}$$

To show that (5.20) implies (5.19) we let

$$h_1(t) = \int_{(t,\infty)} (1 - t/r)^{n-1} \mathrm{d}G(r).$$

Then from the proof given above, we have

$$\int_{(0,\infty)} [h_1(s/x) e^{-x} x^{n-1} / \Gamma(n)] \mathrm{d}x = \int_{(0,\infty)} e^{-s/r} \mathrm{d}G(r).$$

Thus

$$\int_{(0,\infty)} h_1(s/x) e^{-x} x^{n-1} \mathrm{d}x = \int_{(0,\infty)} h(s/x) e^{-x} x^{n-1} \mathrm{d}x,$$

which is equivalent to

$$\int_{(0,\infty)} h_1(1/u) u^{n-1} e^{-su} \mathrm{d}u = \int_{(0,\infty)} h(1/u) u^{n-1} e^{-su} \mathrm{d}u.$$

Hence we have $h = h_1$ by the uniqueness of the Laplace transforms.

Consequently, if $\mathbf{z} \in L_n(G) \cap T_n(h)$, then there is a one-to-one correspondence between G and h in (5.20). □

5.3.2 *Order statistics*

Let $\mathbf{z}^* = (z_1^*, \ldots, z_n^*)' = (z_{(1)}, \ldots, z_{(n)})'$ be the order statistics of $\mathbf{z} = (z_1, \ldots, z_n)$, i.e., $z_{(1)} \leqslant \cdots z_{(n)}$ and $w_j = (n - i + 1)(z_{(i)} - z_{(i-1)})$, $i = 1, \ldots, n (z_{(0)} = 0)$ be the **normalized spacings** (NS) of $\mathbf{z}$.

Theorem 5.12
Assume that $\mathbf{z} \in T_n$ has a density $g(\sum_i^n z_i)$, $\mathbf{z} \in R_+^n$. Then

(1) $\mathbf{z}^*$ has a density

$$n! g\left(\sum_1^n z_i^*\right), \qquad 0 \leqslant z_1^* < \cdots < z_n^*; \tag{5.21}$$

(2) $\mathbf{w} = (w_1, \ldots, w_n) \stackrel{d}{=} \mathbf{z}$.

PROOF Let

$$S_{\pi_i} = \{\mathbf{z}: \mathbf{z} \in R_+^n,\ z_{i_1} < \cdots < z_{i_n}\},$$

where $\pi_i = (i_1, \ldots, i_n)$ is a permutation of $(1, \ldots, n)$. Define the transformation ϕ_i in $S_{\pi_i}: \mathbf{z} \to \mathbf{z}^* = (z_{i_1}, \ldots, z_{i_n})$, the Jacobian of which is 1. Hence $\mathbf{z}^*$ has p.d.f.

$$\sum_{\pi_i} g\left(\sum_{j=1}^n z_{i_j}^*\right) = n! g\left(\sum_{i=1}^n z_i^*\right), \qquad 0 \leqslant z_1^* < \cdots < z_n^*,$$

which proves the first assertion. Consider now the transformation

$$w_i = (n - i + 1)(z_i^* - z_{i-1}^*), \qquad i = 2, \ldots, n$$

$$w_1 = n z_1^*,$$

the Jacobian of which is $n!$. Hence $\mathbf{w}$ has the p.d.f. $g(\sum_1^n u_i)$ which implies $\mathbf{w} \stackrel{d}{=} \mathbf{z}$. □

Remark: If $\mathbf{z} \stackrel{d}{=} \mathbf{x}$ possesses i.i.d. exponential components, so do its normalized spacings. This is one of the basic properties of exponential distributions (cf. e.g. Azlarov and Volodin, 1986). In the case when $\mathbf{z} \in R^n$ does not possess a density, the following is valid.

Theorem 5.13
If $\mathbf{z} \in T_n(h)$, then

$$P(z_{(i)} > y) = \sum_{m=n-i+1}^{n} (-1)^{m-n+i-1} \binom{m-1}{n-i} \binom{n}{m} h(my), \qquad y \geqslant 0. \tag{5.22}$$

If the component z_1 of $\mathbf{z}$ possesses a density $f_1(\cdot)$, then $z_{(i)}$ $(1 \leqslant i \leqslant n)$ has a density

$$f_i^*(y) = \sum_{m=n-i+1}^{n} (-1)^{m-n+i-1} \binom{m-1}{n-i} \binom{n}{m} m f_1(my), \qquad y \geqslant 0. \tag{5.23}$$

PROOF Since $z_1, \ldots z_n$ are exchangeable, it follows from David (1981, equation (5.3.13)) that

$$\begin{aligned} P(z_{n-r+1}^* > y) &= \sum_{m=r}^{n} (-1)^{m-r} \binom{m-1}{r-1} \binom{n}{m} P(z_1 > y, \ldots, z_n > y) \\ &= \sum_{m=r}^{n} (-1)^{m-r} \binom{m-1}{r-1} \binom{n}{m} h(my), \end{aligned}$$

which results in (5.22). Differentiating (5.22) with respect to y and noting $f_1(x) = -h'(x)$ we arrive at formula (5.23). □

The above two theorems are due to Fang and Fang (1988), in which the joint distribution of $z_{(i)}$ and $z_{(j)}$ $(i \neq j)$, and the distributions of the range $z_{(n)} - z_{(1)}$ and the midrange $\frac{1}{2}(z_{(1)} + z_{(n)})$ are also presented. See the problems at the end of this chapter for more details.

Utilizing in Theorem 5.12 more substantially the fact that $\mathbf{w} \stackrel{d}{=} \mathbf{z}$ we obtain a characterization of multivariate ℓ_1-norm symmetric distributions in terms of this property.

Theorem 5.14

Suppose that $\mathbf{z}$ is an n-dimensional exchangeable random vector with p.d.f. $f(\cdot)$ which is a continuous function and let $\mathbf{w}$ be the vector of normalized spacings of $\mathbf{z}$. Then $\mathbf{z} \in T_n$ if $\mathbf{z} \stackrel{d}{=} \mathbf{w}$.

This theorem was first proved by Fang and Fang (1989). We need the following lemma.

Lemma 5.2

Suppose that f is a symmetric function on R_+^n which is continuous in the neighbourhood of $\{\mathbf{u}: u_1 = \cdots = u_{n-1} = 0\}$ and satisfies

$$f(u_1, \ldots, u_n) = f\left(u_1/n, \ldots, \sum_{j=1}^{i} u_j/((n-j+1), \ldots, u_1/n + \cdots + u_n) \right). \tag{5.24}$$

Then

$$f(u_1,\ldots,u_n)=f(0,\ldots,0,\|\mathbf{u}\|). \tag{5.25}$$

PROOF Iterating (5.24) m times we obtain

$$f(\mathbf{u})=f\left(\ldots,\sum_{k=1}^{i} A(i,k;m)u_k,\ldots\right), \tag{5.26}$$

where

$$A(i,k;m)=(n-k+1)^{-1}\sum_{j_1=k}^{i}(n-j_1+1)^{-1}$$
$$\times\sum_{j_2=j_1}^{i}(n-j_2+1)^{-1}\cdots\sum_{j_{m-1}=j_{m-2}}^{i}(n-j_{m-1}+1)^{-1}$$

if $k\leqslant i$, and $A(i,k;m)=0$ if $k>i$. $A(i,k;m)$ is the coefficient of u_k of the ith variable after the mth iteration. The following recursive relations equations are valid:

$$A(i,k;m)=A(i-1,k;m)+A(i,k;m-1)/(n-i+1)$$
$$i=2,\ldots,n;\quad k=1,\ldots,i \tag{5.27}$$

$$\begin{aligned}A(n,k;m)&=(n-k+1)^{-1}\sum_{r=k}^{n}A(n,r;m-1)\\&=(n-k+1)^{-1}A(n,k;m-1)\\&\quad+((n-k)/(n-k+1))A(n,k+1;m)\quad k=1,\ldots,n-1.\end{aligned} \tag{5.28}$$

Hence $A(i,k;m)$ is an increasing sequence in i and $A(n,k;m)$ is an increasing sequence in k and m respectively. Also $\{A(i,k;m)\}$ is bounded. We shall denote the limit of $A(i,k;m)$ by $A(i,k)$ provided it exists. We thus have

$$\begin{aligned}A(i,i)&=0,\qquad i=1,\ldots,n-1\\A(n,n;m)&=A(n,n)=1.\end{aligned} \tag{5.29}$$

Let $m\to\infty$ in (5.27). Then the upper limit

$$\overline{\lim_m}\,A(i,k;m)=\overline{\lim_m}\,A(i-1,k;m)+\overline{\lim_m}\,A(i,k;m)/(n-i+1)$$

or

$$\overline{\lim_m}\,A(i,k;m)=((n-i+1)/(n-i))\,\overline{\lim_m}\,A(i-1,k;m),\qquad i\neq n.$$

From (5.29) we have

$$\overline{\lim_m} A(i,k;m) = 0, \qquad i = 1,\ldots,n-1; k = 1,\ldots,i$$

which leads to

$$A(i,k) = 0, \qquad i = 1,\ldots,n-1; k = 1,\ldots,i.$$

Since $A(n,k;m)$ is an increasing bounded sequence, the limit $A(n,k)$ exists. Setting $m \to \infty$ in (5.28) we have

$$A(n,1) = A(n,2) = \cdots = A(n,n) = 1.$$

Finally, let $m \to \infty$ in (5.26); in view of the continuity of f we obtain (5.25). The lemma is thus proved. □

PROOF OF THEOREM 5.14 Only the sufficiency part requires a proof. From the proof of Theorem 5.12 it follows that $\mathbf{w}$ has the p.d.f.

$$f\left(w_1/n,\ldots,\sum_j^i w_j/(n-j+1),\ldots,w_1/n+\cdots+w_n \right).$$

The assumption $\mathbf{w} \stackrel{d}{=} \mathbf{z}$ leads to (5.24). Sufficiency then follows directly from Lemma 5.2. □

Corollary (Fang and Fang, 1989)
Suppose that $\mathbf{z}$ is an n-dimensional exchangeable random vector with p.d.f. $f(\cdot)$ which is a continuous function. Then $\mathbf{z} \in T_n$ if and only if the joint distribution of normalized spacings of (z_1, z_2) is identical with the conditional distribution of (z_1, z_2) given $(z_3,\ldots,z_n)$.

PROOF The necessity is obvious. We now prove sufficiency. Let $h(z_3,\ldots,z_n)$ be the p.d.f. of $(z_3,\ldots,z_n)$. By Theorem 5.14 the conditional distribution of (z_1, z_2) given $(z_3,\ldots,z_n)$ can be expressed as $g(z_1+z_2|z_3,\ldots,z_n)$. Hence we have

$$\begin{aligned} f(z_1,\ldots,z_n) &= g(z_1+z_2|z_3,\ldots,z_n)h(z_3,\ldots,z_n) \\ &= f(0, z_1+z_2, z_3,\ldots,z_n). \end{aligned}$$

By symmetry of f, this identity leads iteratively to

$$f(z_1,\ldots,z_n) = f\left(0,\ldots,0,\sum_{i=1}^n z_i\right) = g\left(\sum_{i=1}^n z_i\right),$$

which shows that $\mathbf{z}$ is a member of T_n. □

5.4 Mixture of exponential distributions

The topic of mixtures of exponential distributions has recently received a substantial coverage in both monographic and periodical literature (see Titterington, Smith and Mokov, 1985, for an up-to-date survey). Our approach is somewhat different and results in a constructive algorithm for generating appropriate families of distributions.

Let $\mathbf{x}$ consist of i.i.d. standard exponential components and let

$$L_{n,\infty} = \{\mathbf{z}:\mathbf{z} = (z_1, \ldots, z_n)' \stackrel{d}{=} r\mathbf{x}, \text{ where } r \geqslant 0 \text{ is a r.v. independent of } \mathbf{x}\}.$$

Evidently $L_{n,\infty}$ is the family of mixtures of exponential distributions and is a sub-family of L_n.

In this section we shall investigate the family $L_{n,\infty}$ and present several equivalent descriptions. To motivate the theory, a few examples are given below.

For $1 \leqslant m < n, n \geqslant 2$, let

$$L_{m,n-m} = \{\mathbf{z}^{(1)}:\mathbf{z}^{(1)} \in R_+^m \text{ and there exists a } \mathbf{z}^{(2)} \text{ such that } (\mathbf{z}^{(1)'}, \mathbf{z}^{(2)'}) \in L_n\}.$$

Denote $\mathbf{z}^{(1)} \in L_{m,n-m}(F)$ if $\mathbf{z}^{(1)} \in L_{m,n-m}$ and is such that $(\mathbf{z}^{(1)'}, \mathbf{z}^{(2)'})' \in L_n(F)$. The following is an analogue of Eaton's (1981) characterization in Section 4.1.

Theorem 5.15
$\mathbf{z} \in L_{m,n-m}(F)$ iff $\mathbf{z}$ has a density of form $g_m(\sum_{k=1}^m z_k)$ on $R_+^m - \{0\}$, where $g_m(\cdot)$ is given by

$$g_m(t) = \frac{\Gamma(n)}{\Gamma(n-m)} \int_t^\infty (r-t)^{n-m-1} r^{-n+1}\, \mathrm{d}F(r), \qquad t > 0. \qquad (5.30)$$

The proof of this theorem is left to the reader. Since

$$L_{n,n+k} \supset L_{n,n+k+1}, \quad k = 1, 2, \ldots,$$

we have

$$\bigcap_{k=1}^{\infty} L_{n,n+k} \subset L_{n,n+k} \subset L_n, \qquad k > 0.$$

The following theorem shows that $\bigcap_{k=1}^\infty L_{n,n+k} = L_{n,\infty}$.

Theorem 5.16
Let $G(\cdot)$ be a c.d.f. on $(0, \infty)$ and $\mathbf{z} \in R_+^n$. Then the following statements

are equivalent:

(1) $\mathbf{z}\in\bigcap_{k=1}^{\infty}L_{n,n+k}$, and $P(z_1=0)=0$.
(2) $\mathbf{z}\in T_n(h)$, where

$$h(t)=\int_0^{\infty}e^{-t/r}\,\mathrm{d}G(r),\qquad t\geqslant 0.$$

(3) $\mathbf{z}$ has a density of form $f(\|\mathbf{z}\|)$, $\mathbf{z}\in R_+^n$, where

$$f(t)=\int_0^{\infty}e^{-t/r}r^{-n}\,\mathrm{d}G(r),\qquad t>0.$$

(4) $\mathbf{z}\stackrel{d}{=}r\mathbf{x}$, where r has a d.f. $G(\cdot)$ and is independent of $\mathbf{x}$.

PROOF Under the assumption of (1) there exist $z_{n+1},\ldots,z_k$ such that $(z_1,\ldots,z_n,z_{n+1},\ldots,z_k)'\in L_k(G_k)$ for all $k>n$. Then

$$h(t)=P(z_1>t)=\int_t^{\infty}(1-t/r)^{k-1}\,\mathrm{d}G_k(r),\qquad t>0,$$

by Theorem 5.4, so that $h(t)$ is a k-times monotone function for all $k>n$. Hence h is also a k-times monotone function for all $k\leqslant n$, $t>0$ (see Williamson, 1956). Moreover, $h(0)=P(z_1>0)=1$. Thus h is completely monotone on $(0,\infty)$ with the representation

$$h(t)=\int_0^{\infty}e^{-st}\,\mathrm{d}F(s),\qquad t>0,$$

where F is a probability distribution (see Feller, 1971, p. 439). Thus (2) is valid. Next, if (4) holds, then

$$\begin{aligned}P(z_1>t_1,\ldots,z_n>t_n)&=\int_0^{\pi}P(rx_1>z_1,\ldots,rx_n>t_n|r)\,\mathrm{d}G(r)\\&=\int_0^{\infty}e^{-\|\mathbf{z}\|/r}\,\mathrm{d}G(r),\qquad \mathbf{z}\in R_+^n,\end{aligned}$$

yielding (2). By derivation, recalling that $P(r=0)=0$, we obtain (3). It thus follows that the statements (2), (3) and (4) are equivalent. Finally, it is trivial to verify that (4) implies (1). □

Similarly to the above, we may set

$$L_\infty=\{\mathbf{z}:\mathbf{z}'=(z_1,z_2,\ldots)\in R^\infty \text{ and } (z_1,\ldots,z_n)'\in L_n \text{ for each } n\geqslant 1\}.$$

The classes $L_{n,\infty}$ can thus be expressed as:

$$L_{n,\infty} = \{\mathbf{z}:\mathbf{z}' = (z_1,\ldots,z_n), \text{ where there exist } z_k, k = n+1, n+2, \ldots \text{ such that } (z_1,\ldots,z_n, z_{n+1},\ldots)\in L_\infty\}.$$

A theorem similar to Theorem 5.16 can now be stated as follows.

Theorem 5.17
Let $G(\cdot)$ be a d.f. on $(0,\infty)$ and as above, $\mathbf{z}' = (z_1, z_2, \ldots), z_k > 0$, $k = 1, 2, \ldots$ Then the following statements are equivalent:

(1) $\mathbf{z}\in L_\infty$.
(2) $(z_1,\ldots,z_n)'\in T_n(h)$, where

$$h(t) = \int_{(0,\infty)} e^{-t/r}\,\mathrm{d}G(r), \qquad t \geqslant 0, \quad \text{for all } n.$$

(3) $(z_1,\ldots,z_n)$ has the density $f(\|\mathbf{z}\|)$ on R^n_+, where

$$f(x) = \int_0^\infty e^{-x/r} r^{-n}\,\mathrm{d}G(r), \qquad x > 0 \quad \text{for all } n.$$

(4) $\mathbf{z} \overset{d}{=} r\mathbf{x}$, where r and $\mathbf{x}$ are independent, $\mathbf{x}' = (x_1, x_2, \ldots)$ with the x_k being i.i.d. $E(1)$ r.v.s.

Example 5.3
Suppose r is a positive r.v. with density $f(r)$ and independent of $\mathbf{x}$. By Theorem 5.16, if $\mathbf{z} \overset{d}{=} r\mathbf{x}$, then $\mathbf{z}\in L_{n,\infty}$ with a density of form $g_2(\sum_{k=1}^n z_k)$, where

$$g_2(x) = \int_0^\infty e^{-x/r} r^{-n} f(r)\,\mathrm{d}r, \qquad x > 0.$$

If $\mathbf{z} \overset{d}{=} r^{-1}\mathbf{x}$, then $\mathbf{z}\in L_{n,\infty}$ with a density of form $g_3(\sum_{k=1}^n z_k)$, where

$$g_3(x) = \int_0^\infty e^{-xr} r^n f(r)\,\mathrm{d}r, \qquad x > 0.$$

The following are some special cases:

1. Let $f(r) = \mathrm{c}r^k I_{(0,a)}(r)$, where k is a nonnegative integer, $a > 0$, $c = (k+1)/a^{k+1}$ and I is the indicator. Then

$$g_2(x) = ce^{-x/a}\int_0^\infty e^{-x/a}\int_0^\infty e^{-xr}(r + a^{-1})^{n-k-2}\,\mathrm{d}r.$$

For $k \leqslant n-2$, $g_2(x)$ can be expressed as

$$g_2(x) = ce^{-x/a} \sum_{j=0}^{n-k-2} \Gamma(n-k-1)a^{-n+k+2+j}x^{-j-1}/\Gamma(n-k-j)$$

and

$$g_3(x) = -(k+1)e^{-xa}x^{-n-k-1} \sum_{j=0}^{n+k} (n+k)^{(j)}a^{n-j-1}x^{n+k-j} + (k+1)\Gamma(n+k+1)a^{-k-1}x^{-n-k-1}.$$

2. Let $r \sim \mathrm{Ga}(p, 1)$, then

$$g_2(x) = \int_0^\infty e^{-(x/r)-r}r^{-n+p-1}\,\mathrm{d}r/\Gamma(p)$$

and

$$g_3(x) = \Gamma(n+p)(x+1)^{-(n+p)}/\Gamma(p).$$

3. Let $r \sim \mathrm{Be}(p, q)$, then

$$g_2(x) = e^{-x} \int_0^\infty e^{-xr}(r+1)^{n-p-q}r^{q-1}\,\mathrm{d}r/B(p, q).$$

For $n \geqslant p+q$, $g_2(x)$ can be expressed as

$$g_2(x) = e^{-x} \sum_{j=0}^{n-p-q} \Gamma(j+q)\binom{n-p-q}{j}x^{-j-q}/B(p, q)$$

and

$$g_3(x) = \int_0^1 e^{-xr}r^{n+p-1}(1-r)^{q-1}\,\mathrm{d}r;$$

for $q \geqslant 1$, $g_3(x)$ can be expressed as

$$g_3(x) = \sum_{j=0}^{q-1}\left[(-1)^{j+1}\binom{q-1}{j}e^{-x}\sum_{k=0}^{j+n+p-1}(j+n+p-1)^{(k)}x^{-1-k} + (-1)^j\binom{q-1}{j}\Gamma(j+n+p)x^{-j-n-p}\right]\Big/ B(p, q).$$

4. Let $r \sim F(p, q)$, then

$$g_2(x) = \left(\frac{p}{q}\right)^{p/2} \int_0^\infty e^{-xy}y^{n+q/2-1}\left(y+\frac{p}{q}\right)^{-(p+q)/2}\mathrm{d}y/B\left(\frac{p}{2}, \frac{q}{2}\right).$$

Using the above technique one can construct a great many symmetric multivariate distributions. The moments of all orders for mixtures of exponential distributions can be calculated using the definition as follows:

$$E(z_1^{m_1}\ldots z_n^{m_n}) = E(r^{m_1+\cdots+m_n}x_1^{m_1}\ldots x_n^{m_n})$$
$$= E(r^m)\prod_{j=1}^{n} E(x_j^{m_j}) = E(r^m)\prod_{j=1}^{n}(m_j!).$$

Thus only $E(r^m)$ needs to be calculated.

5.5 Independence, robustness and characterizations

It follows from Theorem 5.17 that exponential components $\mathbf{x}$ can serve as a base for the subset $L_{n,\infty}$. The relationship of the exponential distribution to the L_n family is the same as that of the normal distribution to spherically symmetric distributions. Analogously to Section 4.3 we shall present in this section some necessary and sufficient conditions for $\mathbf{z}\in L_n$ to possess independent components. One of these conditions is that $\mathbf{z}\overset{d}{=}\mathbf{x}$. Hence these conditions characterize the exponential distribution. Other characterizations of the exponential distribution will also be given by means of conditional distributions and marginal distributions. These results emphasize once again the special position of the exponential distribution in the L_n family. Finally, we shall discuss the robustness of statistics in this family.

Theorem 5.18
Let $\mathbf{z}\in L_n, n\geqslant 2$. Then the following statements are equivalent:

(1) The components of $\mathbf{z}$ are independently distributed as $E(\lambda)$.
(2) For some $k\neq j$, components z_k and z_j are independent.
(3) For some $k, z_k\sim E(\lambda)$.

PROOF To see that (2) implies (3) we let, without loss of generality, z_1 and z_2 be independent. Then by Theorem 5.4,

$$P(z_1>u_1+u_2) = P(z_1>u_1, z_2>u_2) = P(z_1>u_1)P(z_2>u_2)$$
$$= P(z_1>u_1)P(z_2>u_2), \qquad u_1>0, \quad u_2>0.$$

Hence

$$P(z_1>u_1+u_2|z_1>u_1) = P(z_1>u_2).$$

This ensures that $z_1 \sim E(\lambda)$ for some $\lambda > 0$ (see e.g. Patel, Kapadia and Owen, 1976, p. 131; or Galambos and Kotz, 1978).

Next, (3) implies (1), since

$$P(z_1 > u_1, \ldots, z_n > u_n) = P\left(z_k > \sum_{i=1}^{n} u_i\right) = \exp\left(-\lambda \sum_{i=1}^{n} u_i\right),$$

$$\mathbf{z} \in R_+^n.$$

Finally, it is trivial to show that (1) implies (2). □

Theorem 5.19

Let $\mathbf{z} \stackrel{d}{=} r\mathbf{u} \in L_n(G)$ and $\mathbf{z} = (\mathbf{z}^{(1)'}, \mathbf{z}^{(2)'})'$ where $\mathbf{z}^{(1)}$ is m-dimensional, $1 \leqslant m < n, n \geqslant 2$. The following statements are equivalent:

(1) The components of $\mathbf{z}$ are independently distributed as $E(\lambda)$.
(2) The subvectors $\mathbf{z}^{(1)}$ and $\mathbf{z}^{(2)}$ are independent.
(3) The conditional distribution of $\mathbf{z}^{(1)}$ given $\mathbf{z}^{(2)}$ is nondegenerate with one component being exponential.

PROOF Evidently (1) implies (3).

By Theorem 5.18, since (2) implies the independence of $\mathbf{z}^{(1)}$ and $\mathbf{z}^{(2)}$, we have (1).

It remains to show that (3) implies (2). Under the assumption of (3), using the notation in Section 5.2.3 and Theorem 5.7, for each $\mathbf{w} \in R_+^{n-m}$, we have

$$(\mathbf{z}^{(1)} | \mathbf{z}^{(2)} = \mathbf{w}) \stackrel{d}{=} r_{\|\mathbf{w}\|} \mathbf{u}_1 \in L_m(G_{\|\mathbf{w}\|})$$

with one of its components being conditionally distributed as $E(\lambda(\|\mathbf{w}\|))$. Then by Theorem 5.18 all of its components are conditionally i.i.d. $E(\lambda(\|\mathbf{w}\|))$ and

$$G'_{\|\mathbf{w}\|}(r) = \{\exp[-r\lambda(\|\mathbf{w}\|)]\} r^{m-1} \lambda^m(\|\mathbf{w}\|)/\Gamma(m), \qquad r > 0.$$

Since

$$P(\mathbf{z}^{(2)} = \mathbf{0}) = P(r = 0) \leqslant P(\mathbf{z}^{(1)} | \mathbf{z}^{(2)} \text{ is degenerate}) = 0,$$

we shall only consider the case $\|\mathbf{w}\| > 0$, $G(\|\mathbf{w}\|) < 1$. Then

$$G_{\|\mathbf{w}\|}(r) = \frac{\int_{\|\mathbf{w}\|}^{r+\|\mathbf{w}\|} (s - \|\mathbf{w}\|)^{m-1} s^{-n+1} \, dG(s)}{\int_{\|\mathbf{w}\|}^{\infty} (s - \|\mathbf{w}\|)^{m-1} s^{-n+1} \, dG(s)}.$$

Denoting the denominator by $a(\|\mathbf{w}\|)$, we have

$$1 - G(r + \|\mathbf{w}\|)$$

$$= a(\|\mathbf{w}\|)\lambda^m(\|\mathbf{w}\|)e^{\|\mathbf{w}\|\lambda(\|\mathbf{w}\|)}(\Gamma(m))^{-1}\int_{r+\|\mathbf{w}\|}^{\infty} s^{n-1}e^{-s\lambda(\|\mathbf{w}\|)}\,\mathrm{d}s,$$

for all $r > 0$. Hence

$$G'(r) = a(\|\mathbf{w}\|)\lambda^m(\|\mathbf{w}\|)e^{\|\mathbf{w}\|\lambda(\|\mathbf{w}\|)}(\Gamma(m))^{-1}r^{n-1}e^{-r\lambda(\|\mathbf{w}\|)}$$

for all $r > \|\mathbf{w}\|$. This implies that $\lambda(\|\mathbf{w}\|)$ must be a constant, say λ, and consequently assertion (2) is verified. □

Theorem 5.20

Let $\mathbf{z} \in L_n(G)$. Then the z_k are i.i.d. $E(\lambda)$ iff there exists a positive integer q and a positive constant c, such that $\mathbf{z}^{(1)}$ and $\mathbf{z}^{(2)}$ have densities $g_1(\sum_{k=1}^m z_k)$ and $g_2(\sum_{k=m+1}^n z_k)$, respectively, $\mathbf{z} \in R_+^n$, and

$$g_1(x) = cg_2(x), \qquad x \geqslant 0,$$

where $\mathbf{z} = (\mathbf{z}^{(1)\prime}, \mathbf{z}^{(2)\prime})$, $\mathbf{z}^{(1)}$ is m-dimensional, $\mathbf{z}^{(2)}$ is $(m+q)$-dimensional, and $2m + q = n$, $1 \leqslant m < n$.

PROOF The necessity is obvious. We now show the sufficiency. The letter c is used below as a positive constant, not always the same. By the interchangeability of $\mathbf{z}$,

$$g_1\left(\sum_{k=1}^m z_k\right) = \int_{R_+^q} g_2\left(\sum_{k=1}^{m+q} z_k\right)\mathrm{d}z_{m+1}\ldots\mathrm{d}z_{m+q}$$

$$= c\int_{R_+^q} g_1\left(\sum_{k=1}^{m+q} z_k\right)\mathrm{d}z_{m+1}\ldots\mathrm{d}z_{m+q}, \qquad (z_1,\ldots,z_m)' \in R_+^m.$$

That is,

$$g_1(x) = c\int_{R_+^q} g_1\left(x + \sum_{k=1}^q z_k\right)\mathrm{d}z_1\ldots\mathrm{d}z_q, \qquad x \geqslant 0.$$

Hence

$$g_2\left(\sum_{k=1}^{m+q} z_k\right) = c\int_{R_+^q} g_1\left(\sum_{k=1}^{m+2q} z_k\right)\mathrm{d}z_{m+q+1}\ldots\mathrm{d}z_{m+2q}$$

and

$$\int_{R_+^{m+2q}} g_1\left(\sum_{k=1}^{m+2q} z_k\right) \mathrm{d}z_1 \ldots \mathrm{d}z_{m+2q}$$
$$= c \int_{R_+^{m+q}} g_2\left(\sum_{k=1}^{m+q} z_k\right) \mathrm{d}z_1 \ldots \mathrm{d}z_{m+q} \in (0, \infty).$$

By Theorem 5.9, $cg_1(\sum_{k=1}^{m+2q} z_k)$ is the density of some random vector belonging to the L_{n+2q} family with the $(m+q)$-dimensional marginal density $g_2(\sum_{k=1}^{m+q} z_k)$. Thus $\mathbf{z}^{(2)} \in L_{m+q,q}$. Analogously, $\mathbf{z}^{(2)} \in L_{m+q,jg}$, $j = 2, 3, \ldots$, implying that $\mathbf{z}^{(2)} \in L_{m+q,\infty}$. Hence by Theorem 5.16,

$$g_2(x) = \int_0^\infty e^{-x/r} r^{-m-q} \mathrm{d}G_\infty(r), \qquad x > 0$$

and

$$P(z_{m+1} > t) = \int_0^\infty e^{-t/r} \mathrm{d}G_\infty(r), \qquad t \geqslant 0$$

for some $G_\infty(\cdot)$, a d.f. on $(0, \infty)$. Since by Theorem 5.4

$$P(z_1 > t_1, \ldots, z_n > t_n) = P\left(z_{m+1} > \sum_{i=1}^m t_i\right),$$

$$g_1(x) = \int_0^\infty e^{-x/r} r^{-m} \mathrm{d}G_\infty(r), \qquad x > 0.$$

Then

$$\int_0^\infty e^{-x/r} r^{-m} \mathrm{d}G_\infty(r) = c \int_0^\infty e^{-x/r} r^{-m-q} \mathrm{d}G_\infty(r), \qquad x > 0.$$

The uniqueness of Laplace transforms ensures that

$$\int_0^r s^{-m} \mathrm{d}G_\infty(s) = c \int_0^r s^{-m-q} \mathrm{d}G_\infty(s)$$

for all $r > 0$. Consequently, G_∞ is degenerate at some point $\theta > 0$, $g_1(x) = e^{-x/\theta} \theta^{-m}$ and it follows from Theorem 5.18 that the z_k are i.i.d. $E(\theta^{-1})$. □

Now we shall consider the robustness of the statistics of the L_n family. Let

$$L_n^+ = \{\mathbf{z} : \mathbf{z} \stackrel{d}{=} r\mathbf{u} \in L_n, P(r > 0) = 1\}.$$

A statistic $t(\mathbf{z})$ is said to be **robust in** L_n^+ if $t(\mathbf{z}) \stackrel{d}{=} t(\mathbf{w})$ for any $\mathbf{z}, \mathbf{w} \in L_n^+$. In this case, the distribution of $t(\mathbf{z})$ can be obtained by choosing a convenient $\mathbf{z} \in L_n^+$, say the exponential components $\mathbf{x}$. The following theorem is useful in establishing such a robustness and its proof is similar to that of Theorem 2.22.

Theorem 5.21

Let $\mathbf{z}$ be a member of L_n^+. The statistic $t(\mathbf{z})$ is robust in L_n^+ iff it is scale-invariant, $t(a\mathbf{z}) \stackrel{d}{=} t(\mathbf{z})$ for all $a > 0$.

Example 5.4

1. The statistic

$$t(\mathbf{z}) = \sum_{k=1}^{m} z_k \Big/ \sum_{k=1}^{n} z_k$$

is clearly scale-invariant and thus robust in L_n^+. Therefore using the exponential $\mathbf{x}$ we have

$$t(\mathbf{z}) \stackrel{d}{=} \sum_{k=1}^{m} x_k \Big/ \sum_{k=1}^{n} x_k \sim \mathrm{Be}\,(m, n-m).$$

2. Denote the order statistics of $\mathbf{z} \in L_n^+$ by $z_{(1)} \leqslant z_{(2)} \leqslant \cdots \leqslant z_{(n)}$. The following statistics, useful for detecting and handling outliers, are robust in L_n^+ (cf. Fisher, 1929; Epstein, 1960; Dixon, 1950).

Fisher's type

$$t_1(\mathbf{z}) = \frac{z_{(n)}}{\sum_{i=1}^{n} z_i}, \qquad t_2(\mathbf{z}) = \frac{z_{(1)}}{\sum_{i=1}^{n} z_i}$$

Epstein's type

$$t_3(\mathbf{z}) = \frac{(n-1)(z_{(n)} - z_{(n-1)})}{\sum_{i=1}^{n-1} z_{(i)} + z_{(n-1)}}, \qquad t_4(\mathbf{z}) = \frac{n(n-1)z_{(1)}}{\sum_{i=1}^{n} z_{(i)} - nz_{(1)}},$$

$$t_5(\mathbf{z}) = \frac{n(r-1)z_{(1)}}{\sum_{i=1}^{r-1} z_{(i)} + (n-r+1)z_{(r)} + (1-n)z_{(1)}} \qquad (1 \leqslant r < n-1).$$

It is evident from Definition 2.2 that the elliptically symmetric distributions can be defined by linear random representations of spherically symmetric distributions. This observation naturally leads us to a consideration of families of distributions involving linear random representations of L_n distributions; this topic is discussed in succeeding problems below. For more details, see Fang and Fang (1987a, 1987b). In the conclusion of this chapter we present some problems and recommend that the reader verify these statements.

Problems

5.1 Verify the formulas (5.8), (5.9) and (5.10) in detail.

5.2 Let $\mathbf{u} = (u_1, \ldots, u_n)' \sim v_n$. Then $\mathbf{u}^{(1)} = (u_1, \ldots, u_m)' \in L_m$ $(1 \leqslant m < n)$ and $\mathbf{u}^{(1)} = r_m \mathbf{u}_1$, where $r_m \geqslant 0$ is independent of $\mathbf{u}_1 \sim v_m$.

(i) Derive the p.d.f. of r_m.
(ii) Derive the c.f. of $\mathbf{u}^{(1)}$.
(iii) Derive the conditional distribution of $\mathbf{u}^{(1)}$ given $u_{m+1}, \ldots, u_n$.

5.3 Verify that a necessary and sufficient condition for a nonnegative measurable function $g(\cdot)$ on $(0, \infty)$ to determine a $\mathbf{z} \in L_n$ with the density $C_n g(\|\mathbf{z}\|)$ is that

$$\int_0^\infty x^{n-1} g(x)\,\mathrm{d}x < \infty.$$

5.4 If $\mathbf{z} \in D_{n,\infty}(G)$ and $z_{(1)} \leqslant \cdots \leqslant z_{(n)}$ are its order statistics, show that $z_{(i)}$ possesses the density

$$f_i^*(y) = \frac{n!}{(i-1)!(n-i)!} \int_0^\infty e^{-(n-i+1)y/r}(1 - e^{-y/r})^{i-1} r^{-1}\,\mathrm{d}G(r), \qquad y > 0.$$

In particular, show that for $G(r) = 1 - e^{-r}$,

$$f_i^*(y) = \frac{n!}{(i-1)!(n-i)!} e^{-(n-i+1)y}(1 - e^{-y})^{i-1}, \qquad y > 0.$$

5.5 Let $\mathbf{z} \in T_n(h)$, $1 \leqslant r < s \leqslant n$. Verify that

$$P(z_{(r)} > a, z_{(s)} > b) = \begin{cases} P(z_{(s)} > b), & \text{if } a < 0, \\ P(z_{(r)} > a), & \text{if } a \geqslant b \quad \text{or} \quad b \leqslant 0. \end{cases}$$

Furthermore, if $0 \leqslant a < b$ show that

$$P(z_{(r)} > a, z_{(s)} > b) = \sum_{m=n-r+1}^{n} \sum_{k=n-s+1}^{m} \binom{n}{n-m, m-k, k} \times \sum_{i=0}^{n-k} (-1)^i \sum_{l=0}^{i} \binom{n-m}{l}\binom{m-k}{i-l} \times h((2l+m-k-i)a + (i-l+k)b).$$

(In particular, we have

$$P(z_{(1)} > a, z_{(n)} > b) = \sum_{k=1}^{n} \binom{n}{k} \sum_{i=0}^{n-k} (-1)^i \binom{n-k}{i} \times h((n-k-i)a + (i+k)b).)$$

5.6 Let $\mathbf{z} \in T_n(h)$ and z_1 possess a density $f_1(\cdot)$. Prove that the density of the range $w = z_{(n)} - z_{(1)}$ is:

$$f(w) = \sum_{k=1}^{n-1} \sum_{i=0}^{n-k-1} (-1)^i \frac{(n-1)!(i+k)}{k!(n-k-i-1)!i!} f_1((i+k)w), \quad w > 0.$$

5.7 Let $\mathbf{z} \stackrel{d}{=} r\mathbf{u} \in T_n$. Show that

$$E(z_{(j)}^k) = E(r^k) \sum_{k_1 + \cdots + k_j} \prod_{i=1}^{j} (n-i+1)^{-k_i} \binom{n+k-1}{\text{k}}^{-1}$$

or

$$E(z_{(j)}^k) = E(r^k) \sum_{m=n-j+1}^{n} (-1)^{m-n+j-1} \times \binom{m-1}{n-j}\binom{n}{m} m^{-k} \binom{n+k-1}{k}^{-1}$$

provided that $E(r^k) < \infty$.

5.8 Let

$$S_n(\boldsymbol{\mu}, \mathbf{A}, G) = \{\mathbf{w} : \mathbf{w} \stackrel{d}{=} \boldsymbol{\mu} + \mathbf{A}\mathbf{z}, \mathbf{z} \in L_n(G), \mathbf{A} = \operatorname{diag}(a_1, \ldots, a_n) > 0\},$$

and let $\mathbf{w} \in S_n(\boldsymbol{\mu}, \mathbf{A}, G)$. Verify the following results:

(i)

$$P(w_1 > b_1, \ldots, w_n > b_n) = \begin{cases} 1, & \text{if } b_i < \mu_i, \quad i = 1, \ldots, n \\ \displaystyle\int_{(\alpha_n, \infty)} (1 - \alpha_n/r)^{n-1}\, \mathrm{d}G(r), & \text{otherwise,} \end{cases}$$

where

$$\alpha_n = \sum_{k=1}^{n} \max\left(\frac{b_k - \mu_k}{a_k}, 0\right);$$

(ii) the c.f. of $\mathbf{w}$ is

$$Ee^{i\mathbf{t}'\mathbf{w}} = e^{i\mathbf{t}'\boldsymbol{\mu}} \int_0^\infty \phi_0(r\mathbf{A}\mathbf{t})\,\mathrm{d}G(r),$$

where ϕ_0 is the c.f. of v_n;

(iii) if $\mathbf{z}$ has a density $f(\|\mathbf{z}\|)$ over $R_+^n - \{0\}$, then the density of $\mathbf{w}$ is

$$|\mathbf{A}|^{-1} f(\|\mathbf{A}^{-1}(\mathbf{w}-\boldsymbol{\mu})\|);$$

(iv) if $\mathbf{w}, \boldsymbol{\mu}, \mathbf{z}$ and $\mathbf{A}$ are partitioned into

$$\mathbf{w} = \begin{pmatrix} \mathbf{w}_{(1)} \\ \mathbf{w}_{(2)} \end{pmatrix}, \quad \boldsymbol{\mu} = \begin{pmatrix} \boldsymbol{\mu}_{(1)} \\ \boldsymbol{\mu}_{(2)} \end{pmatrix}, \quad \mathbf{z} = \begin{pmatrix} \mathbf{z}_{(1)} \\ \mathbf{z}_{(2)} \end{pmatrix}, \quad \mathbf{A} = \begin{pmatrix} \mathbf{A}_{11} & \mathbf{0} \\ \mathbf{0} & \mathbf{A}_{22} \end{pmatrix},$$

where $\mathbf{w}_{(1)}, \boldsymbol{\mu}_{(1)}, \mathbf{z}_{(1)}: m \times 1$ and $\mathbf{A}_{11}: m \times m, m < n$, then

$$\mathbf{w}_{(1)} \in S_m(\boldsymbol{\mu}_{(1)}, \mathbf{A}_{11}, G_m),$$

where G_m has a density

$$g_m(x) = (B(m, n-m))^{-1} \int_x^\infty y^{m-1}(r-y)^{n-m-1} r^{-n+1}\,\mathrm{d}G(r),$$

$$x > 0.$$

CHAPTER 6

Multivariate Liouville distributions

In older textbooks on differential and integral calculus (e.g. Edwards, 1922, p. 160; Fichtenholz, 1967, p. 391), the Liouville multiple integral over the positive orthant R^n_+ is derived as an extension of the Dirichlet integral (cf. Section 1.4). It appears to the authors of this book that Marshall and Olkin (1979) were the first to use this integral to define what they called the Liouville–Dirichlet distributions. They show that the expectation operator using these distributions preserves the Schur-convexity property for functions on R^n_+. Sivazlian (1981) gives some results on marginal distributions and transformation properties concerning the Liouville distributions. Anderson and Fang (1982, 1987) study a subclass of Liouville distributions arising from the distributions of quadratic forms. Gupta and Richards (1987), armed with the Weyl fractional integral and Deny's theorem in measure theory on locally compact groups, give a more comprehensive treatment of the multivariate Liouville distribution and also extend some results to the matrix analogues. All the authors mentioned above directly employ the Liouville integral in their approach. In this chapter, however, we shall introduce the Liouville distributions following the approach that emerged in Chapter 2 and was used in Chapter 5: a uniform base with a constraint (here a Dirichlet base) multiplied by a positive generating variate. Some unpublished results of K.W. Ng are included.

6.1 Definitions and properties

The Dirichlet distribution with full parameter $\boldsymbol{\alpha}' = (\alpha_1, \ldots, \alpha_n)$, as discussed in Chapter 1, can be represented either as a distribution on the hyperplane $B_n = \{(y_1, \ldots, y_n)\colon \sum_{i=1}^n y_i = 1\}$ in R^n_+, or as a distribution inside the simplex $A_{n-1} = \{(y_1, \ldots, y_{n-1})\colon \sum_{i=1}^n y_i \leqslant 1\}$ in R^{n-1}_+. For convenience we shall write $\mathbf{y} \sim D_n(\boldsymbol{\alpha})$ on B_n and

$\mathbf{y} \sim D_{n-1}(\alpha_1, \ldots, \alpha_{n-1};\ \alpha_n)$ in A_{n-1} when there is the need to distinguish the two representations.

Definition 6.1

An $n \times 1$ vector $\mathbf{x}$ in R^n_+ is said to have a **Liouville distribution** if $\mathbf{x} \stackrel{d}{=} r\mathbf{y}$, where $\mathbf{y} \sim D_n(\boldsymbol{\alpha})$ on B_n and r is an independent r.v. with c.d.f. F; in symbols, $\mathbf{x} \sim L_n(\boldsymbol{\alpha}; F)$. We shall call $\mathbf{y}$ the **Dirichlet base**, $\boldsymbol{\alpha}$ the **Dirichlet parameter**, r the **generating variate**, and F the **generating c.d.f.** of the Liouville distribution. When r has a density function $f(\cdot)$, we shall call f the **generating density**.

It is clear that $\mathbf{x} \sim L_n(\boldsymbol{\alpha}; F)$ if and only if $(x_1/\Sigma x_i, \ldots, x_n/\Sigma x_i) \sim D_n(\boldsymbol{\alpha})$ on B_n and is independent of the total size Σx_i, which plays the role of r. In general, r need not be an absolutely continuous r.v. and a Liouville distribution may not have a density with respect to the volume measure in R^n_+. As we shall see in Section 6.3, however, all marginal distributions will have density functions if $\mathbf{x} \sim L_n(\boldsymbol{\alpha}; F)$, even if r does not have one.

Theorem 6.1

A Liouville distribution $L_n(\boldsymbol{\alpha}; F)$ has a generating density $f(\cdot)$ if and only if the distribution itself has density

$$\prod_{i=1}^{n} \frac{x_i^{\alpha_i - 1}}{\Gamma(\alpha_i)} \frac{\Gamma(\alpha^*)}{\left(\sum_{i=1}^{n} x_i\right)^{\alpha^* - 1}} f\left(\sum_{i=1}^{n} x_i\right), \qquad \alpha^* = \sum_{i=1}^{n} \alpha_i. \tag{6.1}$$

This density is defined in the simplex $\{(x_1, \ldots, x_n): \sum_{i=1}^{n} x_i \leqslant a\}$ if and only if f is defined in the interval $(0, a)$.

PROOF If r has density $f(\cdot)$ and $\mathbf{y} \sim D_n(\boldsymbol{\alpha})$, the joint density of $y_1, \ldots, y_{n-1}$, and r is

$$\frac{\Gamma(\alpha^*)}{\prod_{i=1}^{n} \Gamma(\alpha_i)} \prod_{i=1}^{n-1} y_i^{\alpha_i - 1} \left(1 - \sum_{i=1}^{n-1} y_i\right)^{\alpha_{n-1}} f(r).$$

Making the transformation

$$\begin{cases} r = \sum_{i=1}^{n} x_i, \\ y_j = x_j \Big/ \sum_{i=1}^{n} x_i, \qquad j = 1, 2, \ldots, n-1, \end{cases}$$

which has Jacobian r^{1-n}, we obtain the p.d.f. of $\mathbf{x} \stackrel{d}{=} r\mathbf{y}$ as given in (6.1). The converse is implied by the inverse of the transformation. The statement about the domain of the density is obvious. □

Note that the p.d.f. (6.1) of $\mathbf{x}$ can be written as

$$\prod_{i=1}^{n} \frac{x_i^{\alpha_i - 1}}{\Gamma(\alpha_i)} g\left(\sum_{i=1}^{n} x_i\right), \tag{6.2}$$

where $g(\cdot)$ is related to the generating density $f(\cdot)$ by the equation

$$g(t) = \frac{\Gamma(\alpha^*)}{t^{\alpha^* - 1}} f(t), \qquad t > 0. \tag{6.3}$$

We shall call $g(\cdot)$ the **density generator** of a Liouville distribution. Note the difference between a density generator and a generating density (cf. Definition 6.1). The following result is a direct consequence of (6.3) and the fact that $f(\cdot)$ is a p.d.f.

Theorem 6.2
A function $g(\cdot)$ on $(0, \infty)$ is the density generator of a Liouville distribution with Dirichlet parameter $(\alpha_1, \ldots, \alpha_n)$ if and only if

$$\int_0^\infty \frac{t^{\alpha^* - 1}}{\Gamma(\alpha^*)} g(t)\,\mathrm{d}t = 1, \tag{6.4}$$

where $\alpha^* = \sum_{i=1}^n \alpha_i$. In this case, the density of the Liouville distribution is given by (6.2) and the generating density $f(\cdot)$ has the univariate Liouville $L_1(\alpha^*; g)$ form as given in (6.3).

From this theorem we see that any function $h(\cdot)$ on $(0, \infty)$ satisfying the condition

$$\int_0^\infty t^{\alpha^* - 1} h(t)\,\mathrm{d}t = C < \infty \tag{6.5}$$

can be made a density generator with a constant adjustment,

$$g(t) = C^{-1}\Gamma(\alpha^*)h(t). \tag{6.6}$$

Besides, using (6.4) and the fact that (6.2) is a p.d.f. we obtain as a by-product the Liouville integral mentioned at the beginning of this chapter,

$$\int_0^\infty \cdots \int_0^\infty g\left(\sum_{i=1}^n x_i\right) \prod_{i=1}^n \frac{x_i^{\alpha_i - 1}}{\Gamma(\alpha_i)} \mathrm{d}x_i = \int_0^\infty \frac{t^{\alpha^* - 1}}{\Gamma(\alpha^*)} g(t)\,\mathrm{d}t, \qquad \alpha^* = \sum_{i=1}^n \alpha_i, \tag{6.7}$$

where the multiple integral on the left is taken inside the simplex $\{\mathbf{x}: \sum_{i=1}^{n} x_i \leqslant a\}$ when the function $g(\cdot)$ is restricted to the finite interval $(0, a)$. Since any nonzero multiplicative constant can be absorbed into the function g, the identity holds as long as either integral is finite. In fact, the name 'multivariate Liouville distribution' is motivated by this integral which is Joseph Liouville's (1809–82) generalization of P.G.L. Dirichlet's integral for the case $g(t) = (1-t)^{a_{n+1}-1}$.

We shall use the notation $L_n(\boldsymbol{\alpha}; f)$ for a Liouville distribution whose generating density is f, and use $L_n(g; \boldsymbol{\alpha})$ instead when emphasizing the distribution's density generator g. Therefore, an absolutely continuous Liouville distribution is determined uniquely by either f or g. In fact, Gupta and Richards (1987) define the Liouville distribution essentially using (6.2) in their pivotal paper in the field. If $g(\cdot)$ has an unbounded domain, the Liouville distribution is said to be of the **first kind**; otherwise, it is of the **second kind**. The density from (6.2) is convenient for quickly identifying a Liouville distribution, while (6.1) is handy for constructing a Liouville distribution using a function $f(\cdot)$ which integrates to unity over $(0, \infty)$.

In analysing so-called 'compositional data' (multivariate observations confined in a simplex), scientists have been handicapped by the lack of known distributions to describe various patterns of variability. Aitchison (1986) gives an excellent account of the difficulties in dealing with compositional data and of the inadequacy of the Dirichlet distribution as a model. In the same book, he offers a clever resolution by transforming the simplex A_n to R^n by the log ratio and other transformations, so that the existing battery of multinormal methods can be readily employed. It has been well received by scientists, especially geologists and biomedical researchers. The constructive formula (6.1), with generating density f restricted to $(0, 1)$, provides yet another route for modelling the compositional data. For a discussion, see Ng (1988a).

Before ending this section we collect some properties of the Liouville distribution which are of immediate consequences in connection with the Dirichlet distributions. Consider first the mixed moments for $\mathbf{x} \stackrel{d}{=} r\mathbf{y}$. The independence between r and $\mathbf{y}$ implies

$$E\left(\prod_{i=1}^{n} x_i^{m_i}\right) = E\left(\prod_{i=1}^{n} (ry_i)^{m_i}\right) = E(r^{\sum m_i})E\left(\prod_{i=1}^{n} y_i^{m_i}\right).$$

Using Theorem 1.3 we obtain the following result.

Theorem 6.3
Let $\mathbf{x} \sim L_n(\boldsymbol{\alpha}; F)$. The mixed moments of $\mathbf{x}$ are given by

$$E\left(\prod_{i=1}^{n} x_i^{m_i}\right) = \mu_m \prod_{i=1}^{n} \alpha_i^{[m_i]}/\alpha^{[m]}, \tag{6.8}$$

where $m = \sum_{i=1}^n m_i$, $\alpha = \sum_{i=1}^n \alpha_i$, μ_m is the mth raw moment of the generating variate, and the power with superscript $[\cdot]$ denotes the ascending factorial, e.g. $\alpha^{[m]} = \alpha(\alpha +)\cdots(\alpha + m - 1)$. In particular, we have

$$E(x_i) = \mu_1 \frac{\alpha_i}{\alpha}, \qquad \operatorname{Var}(x_i) = \frac{\alpha_i}{\alpha}\left(\mu_2 \frac{\alpha_i + 1}{\alpha + 1} - \mu_1^2 \frac{\alpha_i}{\alpha}\right), \tag{6.9}$$

$$\operatorname{Cov}(x_i, x_j) = \frac{\alpha_i \alpha_j}{\alpha}\left(\frac{\mu_2}{\alpha + 1} - \frac{\mu_1^2}{\alpha}\right). \tag{6.10}$$

The following stochastic representations are based on a recursive property of Dirichlet distributions discussed in the next section. Note that part (iii) of Theorem 6.4 is a direct consequence of part (i).

Theorem 6.4
If $\mathbf{x} \sim L_n(\boldsymbol{\alpha},; F)$, then $\mathbf{x}$ can be equivalently represented in the following ways:

(i) $\mathbf{x}' \stackrel{d}{=} r(\prod_{i=1}^{n-1} w_i, (1 - w_1)\prod_{i=2}^{n-1} w_i, \ldots, 1 - w_{n-1})$, where $w_1, \ldots, w_{n-1}$, and r are mutually independent, $w_i \sim \mathrm{Be}(\sum_{j=1}^{i} \alpha_j, \alpha_{i+1})$;
(ii) $\mathbf{x}' \stackrel{d}{=} r(w_1, w_2(1 - w_1), \ldots, w_{n-1}\prod_{i=1}^{n-2}(1 - w_i), \prod_{i=1}^{n-1}(1 - w_i))$, where $w_1, \ldots, w_{n-1}$ and r are mutually independent, $w_i \sim \mathrm{Be}(\alpha_i, \sum_{j=i+1}^{n} \alpha_j)$;
(iii) $\mathbf{x}' \stackrel{d}{=} r(\prod_{i=1}^{n-1}(1 + v_i)^{-1}, w_1\prod_{i=2}^{n-1}(1 + v_i)^{-1}, \ldots, v_{n-1}(1 + v_{n-1})^{-1})$, where $v_1, \ldots, v_{n-1}$ and r are mutually independent, v_i is distributed as inverted beta $IB(\alpha_{i+1}; \sum_{j=1}^{i}\alpha_j)$, i.e. $v_i/(1 + v_i) \sim \mathrm{Be}(\alpha_{i+1}, \sum_{j=1}^{i} \alpha_j)$.

6.2 Examples

The Liouville distributions include several well-known distributions and their generalizations. Examples 6.2 and 6.3 and the marginal properties in Example 6.4 are noted K.W. Ng (1988b).

Example 6.1 (*Dirichlet distribution*)
When $r = 1$ with probability one, the Liouville distribution $L_n(\boldsymbol{\alpha}; F)$ reduces to the Dirichlet distribution $D_n(\boldsymbol{\alpha})$ on B_n.

Example 6.2 (*Beta Liouville distribution*)
Let the generating variate r be distributed as $\mathrm{Be}(\alpha, \beta)$ with generating density

$$f(r) = \frac{1}{B(\alpha, \beta)} r^{\alpha - 1}(1 - r)^{\beta - 1}, \qquad 0 < r < 1 \tag{6.11}$$

and density generator

$$g(r) = \frac{\Gamma(\alpha^*)}{B(\alpha, \beta)} r^{\alpha - \alpha^*}(1 - r)^{\beta - 1}, \qquad 0 < r < 1, \tag{6.12}$$

where $\alpha^* = \sum_{i=1}^n \alpha_i$. Using (6.1) or (6.2), we obtain the joint p.d.f. of $\mathbf{x}$:

$$\frac{\Gamma(\alpha^*)}{B(\alpha, \beta)} \prod_{i=1}^n \frac{x_i^{\alpha_i - 1}}{\Gamma(\alpha_i)} \left(\sum x_i\right)^{\alpha - \alpha^*} \left(1 - \sum x_i\right)^{\beta - 1}. \tag{6.13}$$

To have systematic reference, we shall call (6.11) a **beta Liouville distribution** and write $\mathbf{x} \sim BL_n(\alpha_1, \ldots, \alpha_n; \alpha, \beta)$. This is a parametric generalization of the Dirichlet distributions because when $\alpha = \alpha^*$, $\mathbf{x} \sim D_n(\alpha_1, \ldots, \alpha_n; \beta)$ on A_n. The beta Liouville family is thus a natural model for analysing compositional data (cf. Aitchison, 1986, section 13.1).

Example 6.3 (*Inverted beta Liouville distribution*)
Let the generating variate be $r = \lambda w/(1 - w)$, where w is distributed as $\mathrm{Be}(\alpha, \beta)$. Then r has an **inverted beta** distribution $IB(\alpha; \beta, \lambda)$ with density

$$f(r) = \frac{1}{B(\alpha, \beta)} \cdot \frac{\lambda^\beta r^{\alpha - 1}}{(\lambda + r)^{\alpha + \beta}}, \qquad 0 < r < \infty, \tag{6.14}$$

and density generator

$$g(r) = \frac{\Gamma(\alpha^*)}{B(\alpha, \beta)} \frac{\lambda^\beta r^{\alpha - \alpha^*}}{(\lambda + r)^{\alpha + \beta}}, \qquad 0 < r < \infty, \tag{6.15}$$

where $\alpha^* = \sum_{i=1}^n \alpha_i$. Substituting (6.15) into (6.2), we have the p.d.f. for $\mathbf{x}$:

$$\frac{\Gamma(\alpha^*)}{B(\alpha, \beta)} \prod_{i=1}^n \frac{x_i^{\alpha_i - 1}}{\Gamma(\alpha_i)} \frac{\lambda^\beta (\sum x_i)^{\alpha - \alpha^*}}{(\lambda + \sum x_i)^{\alpha + \beta}} \tag{6.16}$$

When $\alpha = \alpha^*$, this distribution reduces to the so-called **inverted Dirichlet distribution** $ID_n(\alpha_1, \ldots, \alpha_n; \beta, \lambda)$ in R^n_+ (Johnson and Kotz, 1972, p. 239). Following the pattern in Example 6.2, we shall call (6.16) an **inverted beta Liouville distribution** and write $IBL_n(\alpha_1, \ldots, \alpha_n; \alpha, \beta, \lambda)$.

Example 6.4 (Gamma Liouville distribution)
Let r be distributed as $\text{Ga}(\alpha, \lambda)$ with generating density

$$f(r) = \frac{\lambda^\alpha}{\Gamma(\alpha)} r^{\alpha-1} e^{-\lambda r}, \qquad r > 0, \quad \alpha > 0, \quad \lambda > 0, \tag{6.17}$$

and density generator

$$g(r) = \frac{\Gamma(\alpha^*)}{\Gamma(\alpha)} \lambda^\alpha r^{\alpha - \alpha^*} e^{-\lambda r}, \qquad r > 0, \quad \alpha > 0, \quad \lambda > 0, \tag{6.18}$$

where $\alpha^* = \sum_{i=1}^n \alpha_i$. The joint density of $\mathbf{x}$ is given by (6.1):

$$\frac{\lambda^\alpha \Gamma(\alpha^*)}{\Gamma(\alpha) \prod_{i=1}^n \Gamma(\alpha_i)} \left(\sum_{i=1}^n x_i \right)^{\alpha - \alpha^*} \prod_{i=1}^n x_i^{\alpha_i - 1} e^{-\lambda x_i}. \tag{6.19}$$

Letting $\delta = \alpha - \alpha^*$, we can write this density as

$$\frac{\lambda^\delta \Gamma(\alpha^*)}{\Gamma(\delta + \alpha^*)} \left(\sum_{i=1}^n x_i \right)^\delta \prod_{i=1}^n \frac{\lambda^{\alpha_i}}{\Gamma(\alpha_i)} x_i^{\alpha_i - 1} e^{-\lambda x_i}. \tag{6.20}$$

When $\delta = 0$, $\mathbf{x}$ has independent gamma components with common scale parameter λ. For general δ, however, the marginal distributions are not gamma. For an illustration, let $n = 2$, $\lambda = 1$ and $\delta = 1$. Then the marginal density of x_1 is

$$\begin{aligned}
&\frac{\Gamma(\alpha_1 + \alpha_2)}{\Gamma(\alpha_1 + \alpha_2 + 1)} \cdot \frac{x_1^{\alpha_1 - 1} e^{x_1}}{\Gamma(\alpha_1)} \int_0^\infty \frac{(x_1 + x_2) x_2^{\alpha_2 - 1} e^{-x_2}}{\Gamma(\alpha_2)} \mathrm{d}x_2 \\
&= \frac{1}{\alpha_1 + \alpha_2} \frac{x_1^{\alpha_1 - 1} e^{-x_1}}{\Gamma(\alpha_1)} (x_1 + \alpha_2) \\
&= \frac{\alpha_1}{\alpha_1 + \alpha_2} \frac{x_1^{\alpha_1} e^{-x_1}}{\Gamma(\alpha_1 + 1)} + \frac{\alpha_2}{\alpha_1 + \alpha_2} \frac{x_1^{\alpha_1 - 1} e^{-x_1}}{\Gamma(\alpha_1)},
\end{aligned} \tag{6.21}$$

which is a mixture of $\text{Ga}(\alpha_1 + 1, 1)$ and $\text{Ga}(\alpha_1, 1)$ with weights

$\alpha_1/(\alpha_1+\alpha_2)$ and $\alpha_2/(\alpha_1+\alpha_2)$ respectively (see Example 6.10 for more general results). Therefore, the name 'gamma marginals with correlation' suggested by Marshall and Olkin (1979) and the name 'correlated gamma variables' used by Gupta and Richards (1987) may not be the most appropriate. In fact, we shall see in the next section that such Liouville distributions having correlated gamma marginals cannot be constructed. We shall call (6.19) a **gamma Liouville distribution** and write $GL_n(\alpha_1,\ldots,\alpha_n;\alpha,\lambda)$.

Example 6.5 (Symmetric ℓ_1-norm distribution)
Since the uniform distribution on B_n is $D_n(1,\ldots,1)$, the symmetric ℓ_1-norm distribution as discussed in Chapter 5 is a special case of the Liouville distribution.

Example 6.6 (Quadratic forms of spherical distributions)
Let $\mathbf{y} \stackrel{d}{=} r\mathbf{u}^{(n)}$, where r is a positive r.v. independent of $\mathbf{u}^{(n)}$, the uniform base on the surface of the n-dimensional unit sphere. If we partition $\mathbf{y}$ into m parts $\mathbf{y}_1,\ldots,\mathbf{y}_m$ with $n_1,\ldots,n_m$ components respectively, where $n_1+\cdots+n_m=n$, then according to Theorem 2.6 we have

$$\begin{bmatrix} \mathbf{y}_1 \\ \vdots \\ \mathbf{y}_m \end{bmatrix} \stackrel{d}{=} \begin{bmatrix} rd_1\mathbf{u}_1 \\ \vdots \\ rd_m\mathbf{u}_m \end{bmatrix}$$

where $\mathbf{u}_1,\ldots,\mathbf{u}_m$ and $(d_1,\ldots,d_m)$ are mutually independent, $\mathbf{u}_j$ is uniform on the surface of the n_j-dimensional unit sphere, $j=1,\ldots,m$, and $(d_1^2,\ldots,d_m^2)\sim D_m(n_1/2,\ldots,n_m/2)$. Therefore the quadratic forms can be jointly represented as

$$(\mathbf{y}_1'\mathbf{y}_1,\ldots,\mathbf{y}_m'\mathbf{y}_m) \stackrel{d}{=} r^2(d_1^2,\ldots,d_m^2), \tag{6.22}$$

a Liouville distribution with generating variate r^2 and Dirichlet parameters $(n_1/2,\ldots,n_m/2)$. See Anderson and Fang (1987) for more details.

Example 6.7 (Exponential-gamma Liouville distribution)
To give one more distribution suitable for compositional data analysis, we mention the **exponential-gamma Liouville** distribution $EGL_n(\alpha_1,\ldots,\alpha_n;\alpha,\lambda)$ which is constructed by using a generating variate $r=e^{-z}$, where z has a $\mathrm{Ga}(\alpha,\lambda)$ distribution. The 'exponential

gamma' density of r is given by

$$f(r)=\frac{\lambda^{\alpha}}{\Gamma(\alpha)}r^{\lambda-1}\left(\log\frac{1}{r}\right)^{\alpha-1}, \qquad 0<r<1, \quad \lambda>0, \quad \alpha>0 \quad (6.23)$$

and the density generator of $EGL_n(\alpha_1,\ldots,\alpha_n;\alpha,\lambda)$ is

$$g(r)=\frac{\Gamma(\alpha^*)}{\Gamma(\alpha)}\lambda^{\alpha}r^{\alpha-\alpha^*}\left(\log\frac{1}{r}\right)^{\alpha-1}, \qquad 0<r<1, \quad \lambda>0, \quad \alpha>0. \quad (6.24)$$

where $\alpha^*=\sum_{i=1}^n\alpha_i$. Therefore, the density of $\mathbf{x}\sim EGL_n(\alpha_1,\ldots,\alpha_n;\alpha,\lambda)$ is

$$\frac{\Gamma(\alpha^*)}{\Gamma(\alpha)}\lambda^{\alpha}(\sum x_i)^{\alpha-\alpha^*}\left(\log\frac{1}{\sum x_i}\right)^{\alpha-1}\prod_{i=1}^n\frac{x_i^{\alpha_i-1}}{\Gamma(\alpha_i)}. \quad (6.25)$$

We have seen in Example 6.2 that $BL_n(\alpha_1,\ldots,\alpha_n;\sum_{i=1}^n\alpha_i,\alpha_{n+1})$ is $D_n(\alpha_1,\ldots,\alpha_n;\alpha_{n+1})$. This is the recursive property that gives rise to the stochastic representations of Dirichlet distributions in terms of independent beta variables. Let $w_1\sim\mathrm{Be}(\alpha_1,\alpha_2)$ be a Dirichlet base and $w_2\sim\mathrm{Be}(\alpha_1+\alpha_2,\alpha_3)$ be a generating variate. Then the Liouville construction leads to a new Dirichlet base $(w_2w_1, w_2(1-w_1), 1-w_2)\sim D_3(\alpha_1,\alpha_2,\alpha_3)$. Using a new generating variate $w_3\sim\mathrm{Be}(\sum_{i=1}^3\alpha_i;\alpha_4)$, we get a $D_4(\alpha_1,\ldots,\alpha_4)$ distribution for $(w_3w_2w_1, w_3w_2(1-w_1), w_3(1-w_2), 1-w_3)$. Continuing this way, we obtain the stochastic representation of the Dirichlet distribution as given in Theorem 6.4(i). Alternatively, we may let $1-w_i$ play the role of w_i and then renumber the subscripts in reverse order. This route leads to the Dirichlet representation given in Theorem 6.4(ii).

6.3 Marginal distributions

The Liouville distribution has an interesting property that if we fix the Dirichlet parameter, any single marginal distribution identifies the generating distribution and hence the joint distribution.

Theorem 6.5
Let $\mathbf{x}\sim L_n(\boldsymbol{\alpha};F_1)$ and $\mathbf{y}\sim L_n(\boldsymbol{\alpha};F_2)$. If $x_i\stackrel{d}{=}y_i$ for some $i=1,\ldots,n$, then F_1 and F_2 are identical almost surely.

PROOF Consider the separate stochastic representations

$$\mathbf{x}\stackrel{d}{=}r_1\mathbf{w}, \quad \mathbf{y}\stackrel{d}{=}r_2\mathbf{v},$$

where both $\mathbf{w}$ and $\mathbf{v}$ have $D_n(\boldsymbol{\alpha})$ distribution. If $x_1 \stackrel{d}{=} y_1$, we have $r_1 w_1 \stackrel{d}{=} r_2 v_1$. Since $w_1 \stackrel{d}{=} v_1 \sim \text{Be}(\alpha_1, \sum_{j=2}^n \alpha_j)$, which satisfies Lemma 1.2(ii) (cf. Problem 6.4), we have $r_1 \stackrel{d}{=} r_2$. □

Corollary
Assume that $(x_1, \ldots, x_n)$ is distributed as $L_n(\alpha_1, \ldots, \alpha_n; F)$. Then the following are true:

(a) $\mathbf{x}$ is distributed as $D_n(\alpha_1, \ldots, \alpha_n; \alpha_{n+1})$ if and only if for some $i = 1, 2, \ldots, n$, x_i is distributed as $\text{Be}(\alpha_i, \sum_{j \neq i} \alpha_j)$.
(b) $\mathbf{x}$ is distributed as $ID_n(\alpha_1, \ldots, \alpha_n; \beta, \lambda)$ if and only if for some $i = 1, 2, \ldots, n$, x_i is distributed as $IB(\alpha_i; \beta, \lambda)$.
(c) $x_1, x_2, \ldots,$ and x_n are independently distributed as $\text{Ga}(\alpha_i, \lambda)$, $i = 1, 2, \ldots, n$ respectively, if and only if for some $i = 1, \ldots, n$, $x_i \sim \text{Ga}(\alpha_i, \lambda)$.

The last characterization in the corollary dashes our hopes for constructing Liouville distributions having correlated gamma marginal distributions (cf. Example 6.4).

Turning to marginal distributions, we first present the marginal density in terms of F, the generating c.d.f. in the most general case. The expressions in terms of the generating density f and the density generator g (when they exist) then follow as corollaries.

Theorem 6.6
If $\mathbf{x} \sim L_n(\boldsymbol{\alpha}; F)$ and $\mathbf{x}$ has zero probability at the origin, then all the marginal distributions of $\mathbf{x}$ are Liouville distributions having densities. In particular, let $1 \leqslant m \leqslant n$, $\alpha = \sum_{i=1}^n \alpha_i$, $\alpha_1^* = \sum_{i=1}^m \alpha_i$ and $\alpha_2^* = \sum_{i=m+1}^n \alpha_i$, then the density of $(x_1, \ldots, x_m)$ is

$$\prod_{i=1}^m \frac{x_i^{\alpha_i - 1}}{\Gamma(\alpha_i)} \cdot \frac{\Gamma(\alpha)}{\Gamma(\alpha_2^*)} \int_{\sum_{i=1}^n x_i}^{\infty} \frac{\left(r - \sum_{i=1}^m x_i\right)^{\alpha_2^* - 1}}{r^{\alpha - 1}} \mathrm{d}F(r), \tag{6.26}$$

whose density generator is

$$g_m(t) = \frac{\Gamma(\alpha)}{\Gamma(\alpha_2^*)} \int_t^{\infty} \frac{(r - t)^{\alpha_2^* - 1}}{r^{\alpha - 1}} \mathrm{d}F(r), \tag{6.27}$$

and whose generating density is

$$f_m(t) = \frac{\Gamma(\alpha)}{\Gamma(\alpha_1^*)\Gamma(\alpha_2^*)} t^{\alpha_1^* - 1} \int_t^{\infty} \frac{(r - t)^{\alpha_2^* - 1}}{r^{\alpha - 1}} \mathrm{d}F(r). \tag{6.28}$$

PROOF Let $\mathbf{x} \overset{d}{=} r\mathbf{y}$ where $\mathbf{y} \sim D_n(\boldsymbol{\alpha})$ and r has c.d.f. F. The p.d.f. of $(y_1, \ldots, y_m)$ is (cf. Theorem 1.5)

$$\frac{\Gamma(\alpha)}{\Gamma(\alpha_2^*)} \prod_{i=1}^{m} \frac{y_i^{\alpha_i - 1}}{\Gamma(\alpha_i)} \left(1 - \sum_{i=1}^{m} y_i\right)^{\alpha_2^* - 1}.$$

Since $P(r = 0) = 0$, we obtain the multivariate c.d.f. by conditioning:

$$P(x_1 \leqslant t_1, \ldots, x_m \leqslant t_m) = P(y_1 \leqslant t_1/r, \ldots, y_m \leqslant t_m/r)$$

$$= \int_0^\infty \mathrm{d}F(r) \frac{\Gamma(\alpha)}{\Gamma(\alpha_2^*) \prod_{i=1}^{m} \Gamma(\alpha_i)} \int_0^{t_1/r} \cdots \int_0^{t_m/r}$$

$$\times \left(1 - \sum_{i=1}^{m} y_i\right)^{\alpha_2^* - 1} \prod_{i=1}^{m} y_i^{\alpha_i - 1} \mathrm{d}y_i.$$

Taking partial derivatives with respective to $t_1, \ldots, t_m$ at $x_1, \ldots, x_m$ respectively, we have (6.26). By comparing (6.26) and the general form (6.2), we immediately have (6.27). Finally, the general relation (6.3) implies (6.28). □

When the generating c.d.f. $F(\cdot)$ has derivative $f(\cdot)$, the integral in (6.26) can be simplified and hence we have the following results.

Theorem 6.7

Assume further in Theorem 6.6 that the Liouville distribution $L_n(\alpha_1, \ldots, \alpha_n; F)$ has a generating density $f(\cdot)$ and density generator $g(\cdot)$ which are both positive on $(0, a)$, where $0 < a \leqslant \infty$. Let $f_m(\cdot)$ and $g_m(\cdot)$ be respectively the generating density and the density generator for the marginal distribution of $(x_1, \ldots, x_m)$. The following relations hold $(0 < t < a)$:

$$g_m(t) = \frac{\Gamma(\alpha)}{\Gamma(\alpha_2^*)} \int_t^\infty \frac{(r-t)^{\alpha_2^* - 1}}{r^{\alpha - 1}} f(r)\,\mathrm{d}r = \frac{\Gamma(\alpha)}{\Gamma(\alpha_2^*)} \int_0^{a-t} \frac{r^{\alpha_2^* - 1}}{(r+t)^{\alpha - 1}} f(r+t)\,\mathrm{d}r, \tag{6.29}$$

$$g_m(t) = \int_t^\infty \frac{(r-t)^{\alpha_2^* - 1}}{\Gamma(\alpha_2^*)} g(r)\,\mathrm{d}r = \int_0^{a-t} \frac{r^{\alpha_2^* - 1}}{\Gamma(\alpha_2^*)} g(r+t)\,\mathrm{d}r, \tag{6.30}$$

$$f_m(t) = \frac{t^{\alpha_1^* - 1}}{B(\alpha_1^*, \alpha_2^*)} \int_t^\infty \frac{(r-t)^{\alpha_2^* - 1}}{r^{\alpha - 1}} f(r)\,\mathrm{d}r$$

$$= \frac{t^{\alpha_1^*-1}}{B(\alpha_1^*, \alpha_2^*)} \int_0^{a-t} \frac{r^{\alpha_2^*-1}}{(r+t)^{\alpha-1}} f(r+t)\, \mathrm{d}r, \tag{6.31}$$

$$f_m(t) = \frac{t^{\alpha_1^*-1}}{\Gamma(\alpha_1^*)} \int_t^{\infty} \frac{(r-t)^{\alpha_2^*-1}}{\Gamma(\alpha_2^*)} g(r)\, \mathrm{d}r = \frac{t^{\alpha_1^*-1}}{\Gamma(\alpha_1^*)} \int_0^{a-t} \frac{r^{\alpha_2^*-1}}{\Gamma(\alpha_2^*)} g(r+t)\, \mathrm{d}r. \tag{6.32}$$

PROOF Noting that $\mathrm{d}F(r) = f(r)\,\mathrm{d}r$ and changing variable, $u = r - k$, in the integral (6.27), we obtain (6.29). Using the relation (6.3), we get (6.30) from (6.29). We arrive at (6.31) and (6.32) in the same way. □

Johnson and Kotz (1972, p. 305) noted a peculiar property of the Dirichlet distribution that at most one of the component variates can be uniformly distributed over (0, 1). Gupta and Richards (1987, p. 240) attempted to extend this property to the Liouville distributions defined in the simplex A_n. However, a closer look at their proof reveals that the following generalization is valid.

Theorem 6.8
Let $\mathbf{x}$ be distributed as $L_n(g; \alpha_1, \ldots, \alpha_n)$ where the generating variate is confined in an interval $(0, a)$, $a < \infty$. If for some i, x_i is uniformly distributed in $(0, a)$, then any other component x_j such that $\alpha_j < \alpha_i$ cannot be uniformly distributed in $(0, a)$.

PROOF We may, without loss of generality, assume $i = 1$, $j = 2$. Using (6.32) and the assumption that x_1 has constant density in $(0, a)$, we have the following identity function:

$$f_1(t) = \frac{t^{\alpha_1^*-1}}{\Gamma(\alpha_1)} \int_0^{a-t} \frac{r^{\alpha-\alpha_1-1}}{\Gamma(\alpha-\alpha_1)} g(r+t)\, \mathrm{d}r = \frac{1}{a}, \quad 0 < t < a.$$

We first express the marginal density of x_1 as a double integral by means of (6.7) and then evaluate this integral as follows:

$$\begin{aligned} f_2(t) &= \frac{t^{\alpha_2-1}}{\Gamma(\alpha_2)} \int_0^{a-t} \frac{r^{\alpha-\alpha_2-1}}{\Gamma(\alpha-\alpha_2)} g(r+t)\, \mathrm{d}r \\ &= \frac{t^{\alpha_2-1}}{\Gamma(\alpha_2)} \iint_{u+r \leqslant a-t} \frac{u^{\alpha_1-\alpha_2-1}\; r^{\alpha-\alpha_1-1}}{\Gamma(\alpha_1-\alpha_2)\,\Gamma(\alpha-\alpha_1)} g(t+u+r)\, \mathrm{d}u\, \mathrm{d}r \end{aligned}$$

$$= \frac{t^{\alpha_2 - 1}}{\Gamma(\alpha_2)} \int_0^{a-t} \frac{u^{\alpha_1 - \alpha_2 - 1}}{\Gamma(\alpha_1 - \alpha_2)} du \int_0^{a-t-u} \frac{r^{\alpha - \alpha_1 - 1}}{\Gamma(\alpha - \alpha_1)} g(t+u+r)\, dr$$

$$= \frac{1}{a} \frac{t^{\alpha_2 - 1}}{\Gamma(\alpha_2)} \int_0^{a-t} \frac{u^{\alpha_1 - \alpha_2 - 1}}{\Gamma(\alpha_1 - \alpha_2)} \frac{\Gamma(\alpha_1)}{(t+u)^{\alpha_1 - 1}} du$$

$$= \frac{1}{a} \frac{\Gamma(\alpha_1)}{\Gamma(\alpha_2)\Gamma(\alpha_1 - \alpha_2)} \int_0^{(a-t)/t} \frac{w^{\alpha_1 - \alpha_2 - 1}}{(1+w)^{\alpha_1 - 1}} dw, \quad 0 < t < a.$$

Obviously, this density is not a constant function of t, hence we have completed the proof. □

For the special case where $\mathbf{x} \sim D_n(\alpha_1, \ldots, \alpha_n)$, we have $x_i \sim \text{Be}(\alpha_i, \alpha - \alpha_i)$ (Theorem 1.5) for all $i = 1, \ldots, n$. Now if for some i, x_i is uniform on $(0, 1)$, then $\alpha_i = \alpha - \alpha_i = 1$, so that for all $j \neq i$, $\alpha_j < \alpha - \alpha_i = \alpha_i$. According to the above theorem, none of the other component variates can be uniformly distributed on $(0, 1)$.

It is shown in Section 1.4 that the family of Dirichlet distributions is closed under the amalgamation operation. That is, if $\mathbf{x} \sim D_n(\alpha_1, \ldots, \alpha_n; \alpha_{n+1})$ and $v_1 = x_1 + \cdots + x_{n_1}$, $v_2 = x_{n_1+1} + \cdots + x_{n_2}, \ldots, v_k = x_{n_{k+1}} + \cdots + x_n$, then $\mathbf{v} \sim D_k(\alpha_1^*, \ldots, \alpha_k^*; \alpha_{n+1})$, where α_j^* are the amalgamated Dirichlet parameters accordingly. This property can be extended to any family of Liouville distributions having the same generating distribution. In fact, it is part of the following stochastic representation which is an immediate corollary of Theorem 1.4.

Theorem 6.9
Assume $\mathbf{x} \sim L_n(\boldsymbol{\alpha}; F)$ and let $\mathbf{x}' = (\mathbf{x}^{(1)'}, \ldots, \mathbf{x}^{(k)'})$, $\boldsymbol{\alpha}' = (\boldsymbol{\alpha}^{(1)'}, \ldots, \boldsymbol{\alpha}^{(k)'})$, where $\mathbf{x}^{(j)}$ and $\boldsymbol{\alpha}^{(j)}$ have n_j elements, $j = 1, 2, \ldots, k, \sum_{i=1}^k n_j = n$. We have the following stochastic representation:

$$\begin{bmatrix} \mathbf{x}^{(1)} \\ \vdots \\ \mathbf{x}^{(k)} \end{bmatrix} \overset{d}{=} r \begin{bmatrix} w_1 \mathbf{y}^{(1)} \\ \vdots \\ w_k \mathbf{y}^{(k)} \end{bmatrix}, \tag{6.33}$$

where (i) r has c.d.f. F; (ii) $(w_1, \ldots, w_k) \sim D_k(\alpha_1^*, \ldots, \alpha_k^*)$, α_j^* is the ℓ_1-norm of $\boldsymbol{\alpha}^{(j)}$; (iii) $\mathbf{y}^{(j)} \sim D_{n_j}(\boldsymbol{\alpha}^{(j)})$, $j = 1, \ldots, k$; and (iv) the $k + 2$ components $\mathbf{y}^{(1)}, \ldots, \mathbf{y}^{(k)}, \mathbf{w}$ and r are mutually independent.

PROOF Apply Theorem 1.4 to the Dirichlet base $\mathbf{z}$ of $\mathbf{x}$, where $\mathbf{x} \overset{d}{=} r\mathbf{z}$. □

Corollary 1

If $\mathbf{x}$ is distributed as $L_n(\boldsymbol{\alpha}; F)$, then any amalgamation $(v_1, \ldots, v_k)$ of $\mathbf{x}$ is distributed as $L_m(\alpha_1^*, \ldots, \alpha_k^*; F)$, where $(\alpha_1^*, \ldots, \alpha_k^*)$ is the corresponding amalgamation of the Dirichlet parameter $\boldsymbol{\alpha}$.

PROOF For $i = 1, \ldots, k$, v_i is the ℓ_1-norm of $\mathbf{x}^{(i)}$, so that $(v_1, \ldots, v_k) \stackrel{d}{=} r(w_1, \ldots, w_k) \sim L_k(\alpha_1^*, \ldots, \alpha_k^*; F)$. □

Corollary 2

Let r be the generating variate of $\mathbf{x}$, which is distributed as $L_n(\boldsymbol{\alpha}; F)$, and r_m be the generating variate of the marginal distribution of the first m elements of $\mathbf{x}$. Then their moments are related as follows:

$$E(r_m^k) = \frac{\left(\sum_{i=1}^{m} \alpha_i\right)^{[k]}}{\alpha^{[k]}} E(r^k). \tag{6.34}$$

Note that the marginal generating density (6.28) can also be derived from the stochastic representation (6.33) by putting $k = 2$ (cf. Problem 6.5).

As pointed out by Gupta and Richards (1987), the relation (6.30) between a density generator g and its marginal generator g_m is a special case of the Weyl fractional integral transform (for more details, see Rooney, 1972):

$$W^\alpha g(t) = \int_t^a \frac{(r-t)^{\alpha-1}}{\Gamma(\alpha)} g(r)\,\mathrm{d}r = \int_0^{a-t} g(r+t)\,\mathrm{d}r, \tag{6.35}$$

where $0 < t < a \leqslant \infty, \alpha > 0$,

Our development here is self-contained in that the relevant semigroup properties of this integral transform appear as consequences rather than ingredients in the theory. We present simple proofs of these properties.

Theorem 6.10

The Weyl fractional integral transform has the following properties:

(a) If for some $\alpha > 0$, the integral

$$\int_0^a r^{\alpha-1} g(r)\,\mathrm{d}r$$

exists for $0 < a \leqslant \infty$, then $W^{\beta}g$ is well defined in the interval $(0, a)$ for all β such that $0 < \beta \leqslant \alpha$.

(b) The transformation is one-to-one; that is, if for some $\alpha > 0$, $W^{\alpha}g_1(t) = W^{\alpha}g_2(t)$ in an interval $(0, a)$, then $g_1(t) = g_2(t)$ in the same interval.

(c) $W^{\alpha}(W^{\beta}g) = W^{\beta}(W^{\alpha}g) = W^{\alpha+\beta}g$, provided that the last integral exists.

(d) Denote the differential operator by D, then $D(W^{\alpha}g) = -W^{\alpha-1}g$, provided $\alpha - 1 > 0$.

PROOF The condition in (a) implies that $g(\cdot)$, up to a multiplicative constant, is a density generator for any Liouville distribution $L_n(\boldsymbol{\alpha}; g)$ such that $\alpha = \sum_{i=1}^{n} \alpha_i$. Thus, the existence of $W^{\beta}g$ follows from the arbitrariness of α_i and the marginalization (6.30). To show (b), we can view $W^{\alpha}g_1$ and $W^{\alpha}g_2$ as the (marginal) density generators obtained by the same marginalization process applied to two Liouville distributions with identical Dirichlet parameters and density generators g_1 and g_2 respectively. Hence g_1 and g_2 must be equal according to Theorem 6.5. Part (c) follows from the fact that the final marginal distribution of a component is independent of the orders of marginalization processes reducing to this component. Part (d) is obvious once we notice that Wg is simply the negative indefinite integral of g and that $W(W^{\alpha-1}g) = W^{\alpha}$. □

The following special marginal distributions are obtained by K.W. Ng.

Example 6.8 (*Beta Liouville distribution*)
The marginal density of the first component x_1 of a $BL_n(\alpha_1, \ldots, \alpha_n; \alpha, \beta)$ vector (cf. Example 6.2) is given by

$$f(x_1) = \frac{x_1^{\alpha_1 - 1}}{B(\alpha_1, \alpha_2^*)B(\alpha, \beta)} \int_0^{1-x_1} r^{\alpha_2^*-1}(r + x_1)^{\alpha-\alpha^*}(1 - x_1 - r)^{\beta-1}\,dr,$$

which is a formidable integral for general $\alpha - \alpha^*$. However, when $m = \alpha - \alpha^*$ is a nonnegative integer, we may apply the binomial expansion on $(r + x_1)^m$ and obtain

$$f(x_1) = \sum_{k=0}^{m} \binom{m}{k} \frac{B(\alpha_1 + m - k, \alpha_2^* + k)}{B(\alpha_1, \alpha_2^*)} \mathrm{Be}(x_1 | \alpha_1 + m - k, \alpha_2^* + \beta + k), \tag{6.36}$$

where $B(a, b)$ denotes a beta function and $\mathrm{Be}(x|a, b)$ denotes the density function of a $\mathrm{Be}(a, b)$ distribution. Note that the coefficients constitute a **beta-binomial** distribution for variable k (i.e. a beta mixture of binomial distributions). Thus the marginal distribution of x_1 is a beta-binomial mixture of beta distributions when $\alpha - \alpha^*$ is a nonnegative integer. The marginal distribution of x_i can be obtained by substituting α_i for α_1 and $\alpha^* - \alpha_i$ for α_2^* in the above expression because of symmetry.

Example 6.9 (*Inverted beta Liouville distribution*)
Let $\mathbf{x}$ be distributed as $IBL_n(\alpha_1, \ldots, \alpha_n; \alpha, \beta, \lambda)$ which is defined in Example 6.3. Using the technique in the previous example, we can show that for integral values of $m = \alpha - \alpha^* \geqslant 0$ the marginal distribution of x_i is a beta-binomial mixture of inverted beta distributions with the following density:

$$f(x_i) = \sum_{k=0}^{m} \binom{m}{k} \frac{B(\alpha_i + m - k, \alpha^* - \alpha_i + k)}{B(\alpha_i, \alpha^* - \alpha_i)} IB(x_i|\alpha_i + m - k; \beta, \lambda), \tag{6.37}$$

where the inverted beta density is

$$IB(x_i|\alpha_i + m - k; \beta, \lambda) = \frac{\lambda^\beta}{B(\alpha_i + m - k, \beta)} \frac{x_i^{\alpha_i + m - k - 1}}{(\lambda + x_i)^{\alpha_i + m - k + \beta}}. \tag{6.38}$$

Example 6.10 (*Gamma Liouville distribution*)
Let $\mathbf{x}$ be distributed as $GL_n(\alpha_1, \ldots, \alpha_n; \alpha, \lambda)$, which is defined in Example 6.4. When $m = \alpha - \alpha^*$ is a positive integer, the marginal distribution of x_i is a beta-binomial mixture of gamma distributions with the following density:

$$f_i(x_i) = \sum_{k=0}^{m} \binom{m}{k} \frac{B(\alpha_i + m - k, \alpha^* - \alpha_i + k)}{B(\alpha_i, \alpha^* - \alpha_i)} \mathrm{Ga}(x_i|\alpha_i + m - k, \lambda). \tag{6.39}$$

6.4 Conditional distribution

Let us partition $\mathbf{x}$ into two subvectors $\mathbf{x}^{(1)}$ and $\mathbf{x}^{(2)}$. In this section we shall see that if $\mathbf{x}$ is Liouville, the conditional distribution of $\mathbf{x}^{(1)}$ given $\mathbf{x}^{(2)}$ is also a Liouville distribution with the same Dirichlet parameter as for the marginal distribution of $\mathbf{x}^{(1)}$, but having a generating distribution depending on $\mathbf{x}^{(2)}$ only through the ℓ_1-norm $\|\mathbf{x}^{(2)}\|$. When the underlying Liouville distribution has a density function, the conditional density can be obtained by division and thus we have the following result.

Theorem 6.11

Let $(x_1,\ldots,x_n)$ have a Liouville distribution with Dirichlet parameters $(\alpha_1,\ldots,\alpha_n)$, generating density f and density generator g, where $f(\cdot)$ and $g(\cdot)$ are zero outside $(0,a)$, $0<a\leqslant\infty$. The conditional distribution of $(x_1,\ldots,x_m)$ given $(x_{m+1},\ldots,x_n)$, $1\leqslant m<n$, is a Liouville distribution with Dirichlet parameters $(\alpha_1,\ldots,\alpha_m)$ and the following density generator:

$$g_c(t)=\frac{g(t+c)}{W^{\alpha_1^*}g(c)}=\frac{\Gamma(\alpha_1^*)g(t+c)}{\int_0^{a-c}r^{\alpha_1^*-1}g(r+c)\mathrm{d}r},\qquad 0<t<a-c \quad (6.40)$$

where $\alpha_1^*=\sum_{i=1}^m\alpha_i$ and $c=\sum_{i=m+1}^n x_i$. The corresponding conditional generating density is

$$f_c(t)=\frac{t^{\alpha_1^*-1}g(t+c)}{\Gamma(\alpha_1^*)W^{\alpha_1^*}g(c)}=\frac{t^{\alpha_1^*-1}g(t+c)}{\int_0^{a-c}r^{\alpha_1^*-1}g(r+c)\,\mathrm{d}r},\qquad 0<t<a-c \quad (6.41)$$

$$=\frac{t^{\alpha_1^*-1}(t+c)^{1-\alpha}f(t+c)}{\int_0^{a-c}r^{\alpha_1^*-1}(r+c)^{1-\alpha}f(r+c)\,\mathrm{d}r},\qquad 0<t<a-c. \quad (6.42)$$

PROOF The density of $(x_1,\ldots,x_n)$ in terms of g is given by (6.2). Similarly, the density of $(x_{m+1},\ldots,x_n)$ expressed in terms of g is (see (6.30))

$$\prod_{i=m+1}^n\frac{x_i^{\alpha_i-1}}{\Gamma(\alpha_i)}\int_0^{a-c}g(r+c)\frac{r^{\alpha_1^*-1}}{\Gamma(\alpha_1^*)}\mathrm{d}r,\qquad c=\sum_{i=m+1}^n x_i.$$

Dividing (6.2) by this marginal density, we see that the conditional density of $(x_1,\ldots,x_m)$ can be written as

$$\prod_{i=1}^m\frac{x_i}{\Gamma(\alpha_i)}g_c\left(\sum_{i=1}^m x_i\right),$$

where $g_c(\cdot)$ is exactly given by (6.40). Applying (6.3) to the special Liouville distribution $L_m(g_c;\alpha_1,\ldots,\alpha_m)$, we have

$$f_c(t)=\frac{t^{\alpha_1^*-1}}{\Gamma(\alpha_1^*)}g_c(t)=\frac{t^{\alpha_1^*-1}g(t+c)}{\int_0^{a-c}r^{\alpha_1^*-1}g(r+c)\,\mathrm{d}r},\qquad 0<t<a-c,$$

which is just (6.41). Finally, we get (6.42) by substituting (6.3) into (6.41). □

Before extending this result to the general case where the generating variate is not absolutely continuous, we first present a revealing stochastic representation for the conditional distribution.

Lemma 6.1
Assume that $(x_1,\ldots,x_n)\sim L_n(\alpha_1,\ldots,\alpha_n;F)$. Let r be a r.v. having c.d.f. F, and w be independently distributed as $\mathrm{Be}(\alpha_2^*,\alpha_1^*)$, where $\alpha_1^*=\sum_{i=1}^m\alpha_i$ and $\alpha_2^*=\sum_{i=m+1}^n\alpha_i$. For any given $(x_{m+1},\ldots,x_n)$, $c=x_{m+1}+\cdots+x_n$, the conditional distribution of $(x_1,\ldots,x_m)$ is identical with the distribution of the independent product of r_c and $(y_1,\ldots,y_m)$, where $(y_1,\ldots,y_m)$ is $D_m(\alpha_1,\ldots,\alpha_m)$ and r_c is distributed as the conditional distribution of $r-c$ given $rw=cw$ being independently $\mathrm{Be}(\alpha_2^*,\alpha_1^*)$.

PROOF Writing $\mathbf{x}^{(1)}=(x_1,\ldots,x_m)'$ and $\mathbf{x}^{(2)}=(x_{m+1},\ldots,x_n)'$, we apply Theorem 6.9 to this case, $k=2$, getting

$$\begin{bmatrix}\mathbf{x}^{(1)}\\ \mathbf{x}^{(2)}\end{bmatrix}\overset{d}{=}r\begin{bmatrix}(1-w)\mathbf{y}\\ w\mathbf{z}\end{bmatrix},$$

where $r,w,\mathbf{y}$ and $\mathbf{z}$ are mutually independent, r has c.d.f. F, $w\sim\mathrm{Be}(\alpha_2^*,\alpha_1^*)$, $\mathbf{y}\sim D_m(\alpha_1,\ldots,\alpha_m)$ and $\mathbf{z}\sim D_{n-m}(\alpha_{m+1},\ldots,\alpha_n)$. Since $\mathbf{z}$ is independent of all others, the conditional distribution of $\mathbf{x}^{(1)}\overset{d}{=}r(1-w)\mathbf{y}$ depends only on rw, but not on $\mathbf{z}$. Similarly, since $\mathbf{y}$ is independent of (r,w), the conditioning by rw is taken on the first factor inside the independent product $(r-rw)\mathbf{y}$, giving the conclusion as stated in the theorem. □

This lemma points the direction in which we can find the (conditional) c.d.f. of r_c. By definition, if there exists a function $F(\cdot|\cdot)$ of two variables such that

$$P(r(1-w)\leqslant t_1\quad\text{and}\quad rw\leqslant t_2)=\int_0^{t_2}F(t_1|u)h(u)\,du \qquad (6.43)$$

for all $t_1>0$ and $t_2>0$, where $h(u)$ is the marginal density of $u=rw$, then $F(\cdot|c)$ is the conditional c.d.f. of $r(1-w)$ given $rw=c$. Now $h(u)$, being the generating density of $\mathbf{x}^{(2)}$, is given by Theorem 6.6,

$$h(u)=\frac{\Gamma(\alpha)}{\Gamma(\alpha_1^*)\Gamma(\alpha_2^*)}u^{\alpha_2^*-1}\int_u^\infty\frac{(r-u)^{\alpha_1^*-1}}{r^{\alpha-1}}\,dF(r). \qquad (6.44)$$

On the other hand, the region $S = \{(r, w): 0 < r(1-w) \leqslant t_1, 0 < rw \leqslant t_2\}$ is mapped to the region $S' = \{(r, u): 0 < u \leqslant t_2, u < r \leqslant u + t_1\}$ by $r = r$ and $u = rw$, so that when $P(r = 0) = 0$, the left-hand side of (6.43) is

$$\begin{aligned} P(S) &= \iint_S \frac{\Gamma(\alpha)}{\Gamma(\alpha_1^*)\Gamma(\alpha_2^*)} w^{\alpha_2^* - 1}(1-w)^{\alpha_1^* - 1} \, dw \, dF(r) \\ &= \frac{\Gamma(\alpha)}{\Gamma(\alpha_1^*)\Gamma(\alpha_2^*)} \iint_{S'} \frac{u^{\alpha_2^* - 1}(r-u)^{\alpha_1^* - 1}}{r^{\alpha - 1}} \, du \, dF(r) \\ &= \int_0^{t_2} \left\{ \frac{\Gamma(\alpha)}{\Gamma(\alpha_1^*)\Gamma(\alpha_2^*)} u^{\alpha_2^* - 1} \int_u^{u+t_1} \frac{(r-u)^{\alpha_1^* - 1}}{r^{\alpha - 1}} \, dF(r) \right\} du. \end{aligned} \tag{6.45}$$

We are now ready for the following theorem.

Theorem 6.12

Assume that $(x_1, \ldots, x_n)$ is distributed as $L_n(\alpha_1, \ldots, \alpha_n; F)$ with $P(\mathbf{x} = \mathbf{0}) = 0$. Let $c = \sum_{i=m+1}^n x_i$ for any fixed $(x_{m+1}, \ldots, x_n)$, then the conditional distribution of $(x_1, \ldots, x_m)$ given $(x_{m+1}, \ldots, x_n)$ is $L_m(\alpha_1, \ldots, \alpha_m; F_c)$, where the generating c.d.f. $F_c(\cdot)$ is given as

$$F_c(t) = \frac{\int_c^{c+t} (r-c)^{\alpha_1^* - 1} r^{1-\alpha} \, dF(r)}{\int_c^{\infty} (r-c)^{\alpha_1^* - 1} r^{1-\alpha} \, dF(r)}, \qquad t \geqslant 0, c > 0, F(c) < 1; \tag{6.46}$$

when $c = 0$ or $F(c) = 1$, the generating variate $r_c = 0$ with probability one.

PROOF For the usual situations where $c > 0$ and $F(c) < 1$, we compare the right-hand side of (6.43) and the last integral of (6.45), obtaining

$$F(t_1 | u) h(u) = \frac{\Gamma(\alpha)}{\Gamma(\alpha_1^*)\Gamma(\alpha_2^*)} u^{\alpha_2^* - 1} \int_u^{u+t_1} \frac{(r-u)^{\alpha_1^* - 1}}{r^{\alpha - 1}} \, dF(r).$$

Substituting (6.44) into the above equation, we obtain $F(t_1 | u)$. Letting $F_c(t) = F(t | c)$, we get (6.46). Given that $rw = c = 0$, we have $r = 0$ with probability one (i.e. $r_c = 0$ with probability one). Similarly, when $P(r \leqslant c) = F(c) = 1$, the conditional probability is one that $r = c$ (i.e. $r_c = 0$). The proof is completed. □

Since the conditional distribution is also a Liouville distribution, the conditional moments can be computed in the same way as in

Theorem 6.3. The moments of the generating variate r_c are given by

$$E(r_c^k) = \frac{\int_c^\infty (r-c)^{k+\alpha_1^*-1} r^{1-\alpha}\, dF(r)}{\int_c^\infty (r-c)^{\alpha_1^*-1} r^{1-\alpha}\, dF(r)} = \frac{\Gamma(k+\alpha_1^*) W^k g_2(c)}{\Gamma(\alpha_1^*) g_2(c)}, \quad (6.47)$$

where $g_2(\cdot)$ is the density generator of the conditioning components involved. Thus the mixed conditional moments are

$$E\left(\prod_{i=1}^m x_i^{k_i} \,\middle|\, \sum_{i=m+1}^n x_i = t\right) = \prod_{i=1}^m \alpha_i^{[k_i]} W^k g_2(t)/g_2(t), \quad (6.48)$$

where $k = \Sigma k_i$. As shown in the next section, this fact is useful in characterizing special Liouville distributions using regression functions.

Example 6.11 (Dirichlet distribution)
Let $\mathbf{x} \sim D_n(\alpha_1, \ldots, \alpha_n; \beta)$. The marginal density generator of $(x_{m+1}, \ldots, x_n)$ is

$$g_2(r) = \frac{\Gamma(\alpha+\beta)}{\Gamma(\alpha_1^*+\beta)} (1-r)^{\alpha_1^*+\beta-1}, \qquad 0 < r < 1, \alpha = \sum_{i=1}^n \alpha_i. \quad (6.49)$$

The conditional mean of x_k, $1 \leqslant k \leqslant m$, is given by

$$E\left(x_k \,\middle|\, \sum_{i=m+1}^n x_i = t\right) = \frac{\alpha_k \int_t^1 g_2(r)\, dr}{g_2(t)} = \frac{\alpha_k}{\alpha_1^*+\beta}(1-t), \quad (6.50)$$

and the conditional second moments are respectively ($j \neq k$, $1 \leqslant j$, $k \leqslant m$)

$$E\left(x_k^2 \,\middle|\, \sum_{i=m+1}^n x_i = t\right) = \frac{\alpha_k(\alpha_k+1)}{1+\alpha_1^*+\beta}\left(t + \frac{1-t}{\alpha_1^*+\beta}\right), \quad (6.51)$$

$$E\left(x_j x_k \,\middle|\, \sum_{i=m+1}^n x_i = t\right) = \frac{\alpha_j \alpha_k}{1+\alpha_1^*+\beta}\left(t + \frac{1-t}{\alpha_1^*+\beta}\right). \quad (6.52)$$

Example 6.12 (Inverted Dirichlet distribution)
Assuming $\mathbf{x} \sim ID_n(\alpha_1, \ldots, \alpha_n; \beta, \lambda)$, we obtain the marginal density generator of $(x_{m+1}, \ldots, x_n)$,

$$g_2(r) = \frac{\Gamma(\alpha_2^*+\beta)}{\Gamma(\beta)} \frac{\lambda^\beta}{(\lambda+r)^{\alpha_2^*+\beta}}, \qquad 0 < r < \infty. \quad (6.53)$$

For $1 \leqslant k \leqslant m$, the conditional mean of x_k exists when $\alpha_2^* + \beta > 1$,

$$E\left(x_k \middle| \sum_{i=m+1}^{n} x_i = t \right) = \frac{\alpha_k}{\alpha_2^* + \beta - 1}(\lambda + t), \qquad t > 0. \tag{6.54}$$

Similar computations show that the second conditional moments exist when $\alpha_2^* + \beta > 2$ or < 1:

$$E\left(x_k^2 \middle| \sum_{i=m+1}^{n} x_i = t \right) = \frac{\alpha_k(\alpha_k + 1)}{(\alpha_2^* + \beta - 1)(\alpha_2^* + \beta - 2)}(\lambda + t)^2, \tag{6.55}$$

$$E\left(x_j x_k \middle| \sum_{i=m+1}^{n} x_i = t \right) = \frac{\alpha_j \alpha_k}{(\alpha_2^* + \beta - 1)(\alpha_2^* + \beta - 2)}(\lambda + t)^2. \tag{6.56}$$

6.5 Characterizations

In Section 6.3, we have seen that the marginal distributions characterize the classes of Dirichlet, inverted Dirichlet and independent gamma distributions. Here in this section, we shall consider characterizations of these classes using independence structures and regression functions. In this direction, James and Mosimann (1980) presented some characterizations of the Dirichlet class using the concept of neutrality which was introduced by Doksum (1974) in studying neutral random probabilities for tail-free distributions (see also Freedman, 1963). To motivate this concept, we note that the 'awkward' n-dimensional simplex $A_n = \{\mathbf{x}: 0 < x_i < 1, \Sigma x_i < 1\}$ can be transformed to the 'free' n-dimensional cube $I_n = \{\mathbf{y}: 0 < y_i < 1, i = 1, \ldots, n\}$ by the transformation

$$x_1 = y_1, \quad x_i = y_i \prod_{j=1}^{i-1} (1 - y_j), \qquad i = 2, \ldots, n, \tag{6.57}$$

which has the inverse transform

$$y_1 = x_1, \quad y_i = \frac{x_i}{1 - \sum_{j=1}^{i-1} x_j}, \qquad i = 2, \ldots, n \tag{6.58}$$

and Jacobian

$$\left| \frac{\partial(x_1, \ldots, x_n)}{\partial(y_1, \ldots, y_n)} \right| = \prod_{i=1}^{n} (1 - y_i)^{n-i}. \tag{6.59}$$

By induction we have the following relation:

$$\sum_{i=1}^{n} x_i = 1 - \prod_{i=1}^{n} (1 - y_i), \tag{6.60}$$

as if a union of n mutually exclusive events having probabilities x_i could be represented as another union of n mutually independent events having probabilities y_i. Now if $\mathbf{x}$ is a random vector such that $\mathbf{y}$, as defined by (6.57), consists of mutually independent components, we shall call $\mathbf{x}$ and its distribution **completely neutral**. The following theorem says that a Liouville distribution is completely neutral if and only if it is a Dirichlet distribution.

Theorem 6.13

Under the assumption that $\mathbf{x} \sim L_n(g; \alpha_1, \ldots, \alpha_n)$, $\mathbf{x}$ is distributed as $D_n(\alpha_1, \ldots, \alpha_n; \alpha_{n+1})$ for some $\alpha_{n+1} > 0$ if and only if

$$\mathbf{x} \stackrel{d}{=} \left(y_1, y_2(1 - y_1), \ldots, y_n \prod_{i=1}^{n-1} (1 - y_i) \right), \tag{6.61}$$

where $y_1, \ldots, y_n$ are mutually independently distributed in $(0, 1)$.

PROOF If $\mathbf{x}$ is distributed as $D_n(\alpha_1, \ldots, \alpha_{n+1})$, then the density is proportional to

$$x_1^{\alpha_1 - 1} x_2^{\alpha_2 - 1} \ldots x_n^{\alpha_n - 1} \left(1 - \sum_{i=1}^{n} x_i \right)^{\alpha_{n+1} - 1}.$$

Making the transformation (6.57) and using (6.59) and (6.60), we see that the joint density of $(y_1, \ldots. y_n)$ factorizes into independent beta distributions for the y's. Conversely, suppose that (6.61) holds. According to (6.2), the transformed vector $\mathbf{y}$ has a density proportional to

$$y_n^{\alpha_n - 1} \prod_{i=1}^{n-1} y_i^{\alpha_i - 1} (1 - y_i)^{\alpha_{i+1} + \cdots + \alpha_n + n - i} g\left(1 - \prod_{i=1}^{n} (1 - y_i) \right).$$

By mutual independence of y_i, this density should factorize into marginal densities of y_i, so that the last factor in the above expression also factorizes. Therefore, there must exist continuous functions $h_1(\cdot), \ldots, h_n(\cdot)$ such that

$$g\left(1 - \prod_{i=1}^{n} z_i \right) = \prod_{i=1}^{n} h_i(z_i), \qquad 0 < z_i < 1, \quad i = 1, \ldots, n.$$

By symmetry, the functions $h_i(\cdot)$ must be identical. Let $G(\cdot)$ be defined by $G(t) = g(1-t)$, $0 < t < 1$, then we have the following functional equation

$$G\left(\prod_{i=1}^{n} z_i\right) = \prod_{i=1}^{n} h(z_i), \qquad 0 < z_i < 1, \quad i = 1, 2, \ldots, n.$$

Observe that $G(t) = c^{n-1}h(t)$, where $c = \lim_{t\to 1} h(t)$ is the left-hand limit at $t = 1$. Substituting back and rearranging the factors, we have the Cauchy multiplicative functional equation

$$H\left(\prod_{i=1}^{n} z_i\right) = \prod_{i=1}^{n} H(z_i), \qquad 0 < z_i < 1, \quad i = 1, 2, \ldots, n, \tag{6.62}$$

where $H(t) = h(t)/c$, a rescaled function of $h(\cdot)$. The nonzero continuous solution of (6.62) has the form (cf. Problem 6.9)

$$H(t) = t^b, \qquad 0 < t < 1,$$

so that the density generator is given by

$$g(t) = G(1-t) = c^{n-1}h(1-t) = c^n H(1-t) = c^n(1-t)^b. \tag{6.63}$$

Let $\alpha_{n+1} = b + 1$, then the density requirement (6.4) forces that $\alpha_{n+1} > 0$ and $c^n = \Gamma(\alpha_1 + \cdots + \alpha_{n+1})/\Gamma(\alpha_{n+1})$. From (6.2), we see that the distribution of $\mathbf{x}$ is indeed Dirichlet in the n-dimensional simplex. The proof is complete. □

We remark that a proof of this theorem given by Gupta and Richards (1987, p. 251) needs the additional assumption that $\lim_{t\to 0+} g(t) = 1$, which we feel too strong. The above proof is given by K.W. Ng. For the inverted Dirichlet class (cf. Example 6.3), we have a similar result. Indeed, the transformation

$$x_1 = y_1, \quad x_i = y_i \prod_{j=1}^{i-1} (1 + y_i), \quad i = 2, 3, \ldots, n, \tag{6.64}$$

with the inverse transformation

$$y_1 = x_1, \quad y_i = \frac{y_i}{1 + \sum_{j=1}^{i-1} x_j}, \qquad i = 2, \ldots, n, \tag{6.65}$$

and Jacobian

$$\left|\frac{\partial(x_1, \ldots, x_n)}{\partial(y_1, \ldots, y_n)}\right| = \prod_{i=1}^{n} (1 + y_i)^{n-i}, \tag{6.66}$$

maps the positive orthant R^n_+ onto itself. Noting an analogue to (6.60),

$$1 + \sum_{i=1}^{n} x_i = \prod_{i=1}^{n} (1 + y_i), \tag{6.67}$$

and mimicking the proof of Theorem 6.13, we have the following characterization for the inverted Dirichlet class.

Theorem 6.14
Let $(x_1, \ldots, x_n) = \mathbf{x}'$ be distributed as $L_n(g; \alpha_1, \ldots, \alpha_n)$, where the density generator $g(t) > 0$ over $(0, \infty)$. The vector $\mathbf{x}$ is distributed as inverted Dirichlet $ID_n(\alpha_1, \ldots, \alpha_n; \alpha_{n+1})$ for some $\alpha_{n+1} > 0$ if and only if $\mathbf{x}$ has the following stochastic representation:

$$\mathbf{x} \stackrel{d}{=} \left(y_1, y_2(1 + y_1), \ldots, y_n \prod_{i=1}^{n-1} (1 + y_i) \right), \tag{6.68}$$

where $y_1, \ldots, y_n$ are mutually independently distributed over $(0, \infty)$.

We now turn our attention to the joint distributions of independent gamma variables.

Theorem 6.15
Assume that $\mathbf{x}$ is distributed according to $L_n(\alpha_1, \ldots, \alpha_n; F)$ and that $P(\mathbf{x} = \mathbf{0}) = 0$. The following conditions are equivalent:

(a) $x_1, \ldots, x_n$ are independently distributed as $\text{Ga}(\alpha_i, \lambda)$ for some $\lambda > 0$;
(b) $x_1, \ldots, x_n$ are mutually independent;
(c) there exists a pair of independent subvectors of $\mathbf{x}$;
(d) there exist two subvectors of $\mathbf{x}$, the respective sums of whose components are independent.

PROOF We need only show that (d) implies (a). Because of symmetry, we may assume, without loss of generality, that the two subvectors are partitioned as $\mathbf{x}^{(1)} = (x_1, \ldots, x_m)$ and $\mathbf{x}^{(2)} = (x_{m+1}, \ldots, x_{m+k})$, $1 \leqslant m < m + k \leqslant n$, with the complementary subvector being $\mathbf{x}^{(3)} = (x_{m+k+1}, \ldots, x_n)$. Let (y_1, y_2, y_3) be the vector of the sums of the subvectors and $(\alpha_1^*, \alpha_2^*, \alpha_3^*)$ the corresponding amalgamation of the Dirichlet parameters, then (y_1, y_2, y_3) is distributed as $L_3(\alpha_1^*, \alpha_2^*, \alpha_3^*; F)$, according to Corollary 1 of Theorem 6.9. The marginal distribution of (y_1, y_2) is also Liouville $L_2(\alpha_1^*, \alpha_2^*; F^*)$.

That is, $(y_1, y_2) \stackrel{d}{=} r^*(w_1, w_2)$, where $w_1 + w_2 = 1$, w_1 is distributed as $\mathrm{Be}(\alpha_1^*, \alpha_2^*)$, r^* has c.d.f. F^* and is independent of (w_1, w_2). Since $P(\mathbf{x} = \mathbf{0}) = 0$, we see that $y_1/y_2 \stackrel{d}{=} w_1/w_2$ is independent of $y_1 + y_2 = r^*$. In view of a characterization of gamma distributions, as given in Lukacs (1956, p. 208), y_1 and y_2 are independently $\mathrm{Ga}(a_1, \lambda)$ and $\mathrm{Ga}(a_2, \lambda)$ for some $a_1 > 0$, $a_2 > 0$ and $\lambda > 0$, so that $r^* = y_1 + y_2$ has a density $g^*(\cdot)$. Now the distribution of (y_1, y_2) has two equivalent forms, one being $L_2(\alpha_1^*, \alpha_2^*; g^*)$ and one from independence:

$$C_1 y_1^{\alpha_1 - 1} y_2^{\alpha_2 - 1} g^*(y_1 + y_2) = C_2 y_1^{a_1 - 1} y_2^{a_2 - 1} e^{-\lambda(y_1 + y_2)}.$$

The special form of the last factor on the left-hand side dictates that $a_1 = \alpha_1^*$, $a_2 = \alpha_2^*$. In view of the corollary of Theorem 6.5, it is clear that y_1, y_2 and y_3 are mutually independent with $y_i \sim \mathrm{Ga}(\alpha_i^*, \lambda)$, $i = 1, 2, 3$. So $r = y_1 + y_2 + y_3 = \sum_{i=1}^n x_i$, the generating variate, has a $\mathrm{Ga}(\sum_{i=1}^n \alpha_i, \lambda)$ distribution. This implies (a), according to the discussion in Example 6.4, completing the proof. □

The equivalence of the conditions (a), (b) and (c) in Theorem 6.15 is formulated for the absolute continuous case in Gupta and Richards (1987), while condition (d) is essentially due to Anderson and Fang (1987). The proof here is a modified version of the latter.

The above three subclasses of Liouville distributions can also be characterized by means of their regression functions. These results shall be shown below as illustrations of a technique which hinges on the following lemma. We recall that for Liouville distributions a conditional distribution depends on the conditioning variables only through their total.

Lemma 6.2

Let $(x_1, \ldots, x_n)$ have a Liouville distribution and let $h(t)$ be the regression function of x_k on $(x_{m+1}, \ldots, x_n)$, where $1 \leqslant k \leqslant m < n$, $\sum_{i=m+1}^n x_i = t$. Then the density generator $g(\cdot)$ of the subvector $(x_{m+1}, \ldots, x_n)$ is given by

$$g(t) = c \exp\left\{ -\int \frac{h'(t) - \alpha_k}{h(t)} \, \mathrm{d}t \right\}, \tag{6.69}$$

where $c > 0$ is a constant and $h'(t)$ is the derivative of $h(t)$.

PROOF From (6.48) we have

$$h(t) = E\left(x_k \middle| \sum_{i=m+1}^{n} x_i = t \right) = \alpha_k \int_t^\infty g(u)\,\mathrm{d}u / g(t). \tag{6.70}$$

Differentiating (with respect to t) the integral equation

$$\alpha_k \int_t^\infty g(u)\,\mathrm{d}u = h(t)g(t),$$

and rearranging terms, we have

$$\frac{g'(t)}{g(t)} = -\frac{h'(t) + \alpha_k}{h(t)}.$$

Integrating both sides, we obtain (6.69). □

This lemma, coupled with Theorem 6.5, provides a means of characterizing special subclasses among Liouville distributions.

Theorem 6.16
Let $(x_1, \ldots, x_n)$ be distributed as $L_n(\alpha_1, \ldots, \alpha_n; F)$ in R_+^n such that $P(\mathbf{x} = \mathbf{0}) = 0$. Then $x_1, \ldots, x_n$ are independently distributed as $\mathrm{Ga}(\alpha_i, \lambda)$, $i = 1, \ldots, n$, for some $\lambda > 0$ if and only if some regression function of a component is a constant function of the conditioning components.

PROOF If $h(t) = E(x_k | \sum_{i=m+1}^n x_i = t) = b$ for all $t > 0$, we have $b > 0$ since x_k cannot take negative values. Lemma 6.2 gives the density generator for $(x_{m+1}, \ldots, x_n)$:

$$g(t) = ce^{-\lambda t}, \tag{6.71}$$

where $\lambda = \alpha_k / b > 0$. The density generator requirement (6.4) implies $c = \lambda^{\alpha_2^*}$, where $\alpha_2^* = \sum_{i=m+1}^n \alpha_i$. This means (Example 6.4) that $x_{m+1}, \ldots, x_n$ are independent $\mathrm{Ga}(\alpha_i, \lambda)$, $i = m+1, \ldots, n$. Thus the corollary to Theorem 6.5 extends this conclusion for $i = 1, 2, \ldots, n$. The converse is obvious because of independence. □

We remark that Gupta and Richards (1987, p. 247) characterize the independent gamma components using constancy of conditional moments $E(x_k^j | \sum_{i=m+1}^n x_i = t)$ for all j. Their proof involves **complete monotonicity** and a general form of the Hausdorff–Bernstein theorem. The proof presented here, Lemma 6.2, and the following two characterizations are due to K.W. Ng.

Theorem 6.17

Let $\mathbf{x}$ be distributed as $L_n(\alpha_1,\ldots,\alpha_n;F)$ in R_+^n such that $P(\mathbf{x}=\mathbf{0})=0$. Then $\mathbf{x}$ is distributed as inverted Dirichlet $ID_n(\alpha_1,\ldots,\alpha_n;\beta,\lambda)$ for some $\lambda>0$ and $\beta>0$ if and only if some regression function of a component is a linear function of the conditioning components.

PROOF Suppose $h(t)=E(x_k|\sum_{i=m+1}^n x_i=t)=a+bt$, $t>0$. Using Lemma 6.2, we obtain the density generator of $(x_{m+1},\ldots,x_n)$:

$$g(t)=c(\lambda+t)^{-d}, \tag{6.72}$$

where $\lambda=a/b>0$ and $d=(b+\alpha_k)/b>0$. The requirement (6.4) implies that $d=\alpha_2^*+\beta$ for some $\beta>0$ and that $(x_{m+1},\ldots,x_n)$ is distributed as (see Example 6.3) the inverted Dirichlet distribution $ID_{n-m}(\alpha_{m+1},\ldots,\alpha_n;\beta,\lambda)$. This in turn implies that $\mathbf{x}$ is distributed as $ID_n(\alpha_1,\ldots,\alpha_n;\beta,\lambda)$. The converse follows from (6.54). □

Theorem 6.18

Let $\mathbf{x}$ be distributed as $L_n(\alpha_1,\ldots,\alpha_n;F)$ in the simplex A_n. Then $\mathbf{x}$ is distributed as $D_n(\alpha_1,\ldots,\alpha_n;\beta)$ for some $\beta>0$ if and only if some regression function of a component is a linear function of the conditioning components.

PROOF A linear regression function can be expressed as

$$h(t)=E\left(x_k\,\middle|\,\sum_{i=m+1}^n x_i=t\right)=a(1-t)+b, \qquad 0\leqslant t\leqslant 1,$$

where constants a and b are to be determined. As $t\to 1$, the conditional sample space of x_k, namely the interval $[0,1-t]$, shrinks to the origin, and so does the conditional mean of x_k. This implies $b=0$. Furthermore, letting $t\to 0$, we see that

$$a=E(x_k)=\frac{\alpha_k}{\alpha_1^*}\mu,$$

where $\alpha_1^*=\sum_{i=1}^m\alpha_i$ and μ is the mean of the generating variate of $(x_1,\ldots,x_m)$, $0<\mu<1$. According to Lemma 6.2, the density generator of $(x_{m+1},\ldots,x_n)$ is given by

$$g(t)=c(1-t)^{\alpha_k/a-1}=c(1-t)^{\alpha_1^*/\mu-1}.$$

Since $0<\mu<1$, $\alpha_1^*/\mu>\alpha_1^*$, so that $\alpha_1^*/\mu=\alpha_1^*+\beta$ for some $\beta>0$. This means (see Example 6.2) that $(x_{m+1},\ldots,x_n)$ is distributed as

$D_{n-m}(\alpha_{m+1},\ldots,\alpha_n;\alpha_1^*+\beta)$. This marginal distribution identifies $\mathbf{x}\sim D_n(\alpha_1,\ldots,\alpha_n;\beta)$. The converse follows from (6.50). □

6.6 Scale-invariant statistics

In Chapters 2 and 5, we have seen that the distribution of a scale-invariant statistic is not affected by a change of the generating distribution. This is also true for Liouville distributions.

Theorem 6.19

Let $\mathbf{x}$ be distributed as $L_n(\alpha_1,\ldots,\alpha_n;F)$ such that $P(\mathbf{x}=\mathbf{0})=0$. The distribution of a statistic $t(\mathbf{x})$ does not depend on the choice of F if and only if it is true that $t(a\mathbf{x})\stackrel{d}{=}t(\mathbf{x})$ for any $a>0$.

PROOF Let $\mathbf{x}\stackrel{d}{=}r\mathbf{y}$, where r has c.d.f. F and $\mathbf{y}$ is independently distributed as $D_n(\alpha_1,\ldots,\alpha_n)$. For the sufficiency we need only show that $t(\mathbf{x})\stackrel{d}{=}t(\mathbf{y})$. Indeed, the c.f. of $t(\mathbf{x})$ is given by

$$\psi(s)=E(e^{ist(\mathbf{x})})=\int_0^\infty E(e^{ist(r\mathbf{y})}|r)\,\mathrm{d}F(r)$$

$$=\int_0^\infty E(e^{ist(\mathbf{y})})\,\mathrm{d}F(r)=E(e^{ist(\mathbf{y})}),$$

which equals the c.f. of $t(\mathbf{y})$. For the necessity, note that the non-dependence on F means $t(r\mathbf{y})\stackrel{d}{=}t(r_1\mathbf{y})$ for any other generating variate r_1 and hence for $r_1\stackrel{d}{=}ar$, where a is any positive number. □

Corollary

Assume as in Theorem 6.19. Then the joint distribution of $(t_1(\mathbf{x}),\ldots,t_k(\mathbf{x}))$ does not depend on the choice of F if and only if for each i $(i=1,\ldots,k)$ it is true that $t_i(\mathbf{x})\stackrel{d}{=}t_i(a\mathbf{x})$ for all $a>0$.

PROOF The necessity is trivial. For the sufficiency, recall that, in general, a joint distribution is characterized by the set of distributions of all possible linear combinations of the component variables. Now, since each component's distribution does not depend on the choice of F, neither does any linear combination of these components. Thus the result is proved. □

These results give us the convenience of using special Liouville distributions to find the distributions for particular scale-invariant statistics. Also, if our inferences from some special Liouville model (such as independent gamma lifetimes) are based on scale-invariant statistics, they can be extended to all Liouville models with the same set of Dirichlet parameters. In this sense, the scale-invariant statistics are robust. Ratios of linear combinations of components (or ordered components) are typical scale-invariant statistics. In the examples below we shall assume $\mathbf{x} \sim L_n(\alpha_1, \ldots, \alpha_n; F)$ with an arbitrary generating distribution function F.

Example 6.13
The statistic $t(\mathbf{x}) = \sum_{i=1}^{m} x_i / \sum_{i=1}^{n} x_i$ is scale-invariant. We may choose $\mathbf{x} \sim D_n(\alpha_1, \ldots, \alpha_n)$ (the case when $P(\sum_{i=1}^{n} x_i = 1) = 1$) or $x_i \sim \mathrm{Ga}(\alpha_i, 1)$ to show that $t(\mathbf{x})$ is $\mathrm{Be}(\sum_{i=1}^{m} \alpha_i, \sum_{i=m+1}^{n} \alpha_i)$.

Example 6.14 (Predicting a Liouville component)
Consider the nonparametric problem of predicting the lifetime x_n, based on the lifetimes $x_1, \ldots, x_{n-1}$, assuming that $\mathbf{x}$ has a Liouville distribution with known α_i but unknown generating distribution. Being scale-invariant, the statistic $t(\mathbf{x}) = x_n / \sum_{i=1}^{n-1} x_i$ has a known distribution. Thus frequency tolerance intervals can be constructed for x_n based on $\sum_{i=1}^{n-1} x_i$. Taking the convenient choice of independent $\mathrm{Ga}(\alpha_i, 1)$ distributions for the x_i, we can show that (cf. Problem 6.11) $t(\mathbf{x})$ has the inverted beta $IB(\alpha_n; \sum_{i=1}^{n-1} \alpha_i, 1)$ distribution. Let w be a $\mathrm{Be}(\alpha_n, \sum_{i=1}^{n-1} \alpha_i)$ variable and $P(w_1 < w < w_2) = \beta$. Then we have the following prediction statement for x_n:

$$P\left(\frac{w_1}{1 - w_1} \sum_{i=1}^{n-1} x_i < x_n < \frac{w_2}{1 - w_2} \sum_{i=1}^{n-1} x_i\right) = \beta. \tag{6.73}$$

Beta probabilities and quantiles are given in many statistical packages, such as Minitab and SAS.

Example 6.15 (Predicting a Liouville subvector)
A more general problem is finding a joint predictive distribution of the last $n - m$ observations based on the first m observations. According to the corollary to Theorem 6.19, the joint distribution of $(t_{m+1}(\mathbf{x}), \ldots, t_n(\mathbf{x}))$, where $t_j(\mathbf{x}) = x_j / \sum_{i=1}^{m} x_i$, $j = m + 1, \ldots, n$, does not depend on the choice of F. Therefore we may again choose x_i being $\mathrm{Ga}(\alpha_i, 1)$ to show that $(t_{m+1}, \ldots, t_n)$ has an inverted Dirichlet

distribution $ID_{n-m}(\alpha_{m+1},\ldots,\alpha_n;\sum_{i=1}^m \alpha_i, 1)$. Except for special cases (see Example 6.18), computations using this predictive distribution would involve numerical techniques of integration. One can give some bounds for the predictive probabilities by means of the inequalities discussed later (see Example 6.20).

Example 6.16

Here we list some scale-invariant statistics which are useful in model diagnostics and other applications in life testing (cf. part 2 of Example 5.4). Let $x_{(1)} \leqslant x_{(2)} \leqslant \cdots \leqslant x_{(n)}$ be the ordered components of $\mathbf{x}$.

$$t_1(\mathbf{x}) = x_{(n)} \Big/ \sum_{i=1}^n x_i, \quad t_2(\mathbf{x}) = x_{(1)} \Big/ \sum_{i=1}^n x_i,$$

$$t_3(\mathbf{x}) = \frac{(n-1)(x_{(n)} - x_{(n-1)})}{\sum_{i=1}^{n-1} x_{(i)} + x_{(n-1)}}, \quad t_4(\mathbf{x}) = \frac{n(n-1)x_{(1)}}{\sum_{i=1}^n x_i - n x_{(1)}}$$

$$t_5(\mathbf{x}) = \frac{n(m-1)x_{(1)}}{\sum_{i=1}^{m-1} x_{(i)} + (n-m+1)x_{(m)} + (1-n)x_{(1)}}, \qquad 1 \leqslant m < n-1.$$

Note that we may choose $\mathbf{x} \sim D_n(\alpha_1,\ldots,\alpha_n)$, hence $t_1(\mathbf{x})$ is distributed as the maximum component and $t_2(\mathbf{x})$ as the minimum component of a Dirichlet vector. Note also that $t_4(\mathbf{x})$ is an increasing function of $t_2(\mathbf{x})$.

6.7 Survival functions

We now consider the computation of the survival function $S(t_1,\ldots,t_n) = P(x_1 \geqslant t_1,\ldots,x_n \geqslant t)$, assuming $\mathbf{x} \sim L_n(g;\alpha_1,\ldots,\alpha_n)$ with density generator $g(t)$, $0 < t < \infty$. The approach here is different from that in Section 5.2.1, because the survival function of a general Dirichlet base, unlike the uniform base, is part of the problem. Taking the transformation $z_j = \sum_{i=j}^n x_i$, $j = 1,\ldots,n$, we have

$$P(x_1 \geqslant t_1,\ldots,x_n \geqslant t_n)$$
$$= \int_{t_n}^\infty \frac{z_n^{\alpha_n - 1}}{\Gamma(\alpha_n)} \mathrm{d}z_n \int_{t_{n-1}+z_n}^\infty \frac{(z_{n-1} - z_n)^{\alpha_{n-1}-1}}{\Gamma(\alpha_{n-1})} \mathrm{d}z_{n-1}$$
$$\cdots \int_{t_1+z_2}^\infty \frac{(z_1 - z_2)^{\alpha_1 - 1}}{\Gamma(\alpha_1)} g(z_1)\,\mathrm{d}z_1. \qquad (6.74)$$

For general Dirichlet parameters α_i, we need numerical techniques to evaluate this repeated integral. Let us explore an analytic approach by extending the Weyl operator (6.35) as follows:

$$W_\beta^\alpha g(t) = \int_{\beta+t}^\infty \frac{(r-t)^{\alpha-1}}{\Gamma(\alpha)} g(r)\,\mathrm{d}r, \qquad \alpha>0, \quad \beta>0, \quad t>0. \tag{6.75}$$

We see that the survival function can be expressed as a sequence of operations on the density generator, then evaluated at the origin,

$$S(t_1,\ldots,t_n) = W_{t_n}^{\alpha_n}(\ldots(W_{t_1}^{\alpha_1}g))(0). \tag{6.76}$$

Unfortunately, little is known about the integral transform (6.75). We do know some simple properties as given in the following theorem.

Theorem 6.20

The extended Weyl operator (6.75) and the Weyl operator (6.35) have the following relationships:

(a) $W_t g(x) = Wg(t+x)$.

(b) $W_0^\alpha g(x) = W^\alpha g(x)$.

(c) $W_t^m g(x) = \sum_{j=0}^{m-1}(t^j/j!)W^{m-j}g(t+x)$, provided that $m\geqslant 1$ is an integer.

(d) For integers m_1 and m_2,

$$(W_{t_2}^{m_2}(W_{t_1}^{m_1}g))(x) = \sum_{k=0}^{m_2-1}\sum_{j=0}^{m_1-1}\frac{t_1^j}{j!}\frac{t_2^k}{k!}W^{m_1+m_2-j-k}g(t_1+t_2+x).$$

PROOF Parts (a) and (b) are straightforward. For (c) we make use of the binomial expansion

$$\frac{(r-x)^{m-1}}{\Gamma(m)} = \sum_{j=0}^{m-1}\frac{t^j}{j!}\frac{(r-t-x)^{m-1-j}}{(m-1-j)!}$$

so that

$$W_t^m g(x) = \sum_{j=0}^{n-1}\frac{t^j}{j!}\int_{t+x}^\infty \frac{(r-t-x)^{m-j-1}}{\Gamma(m-j)}g(x)\mathrm{d}r.$$

This proves (c). Part (d) follows from the linearity and semigroup properties of the Weyl operator (cf. Theorem 6.9). □

Thus we see that when α_i are integers, the survival function is in a manageable form analytically. For instance, when $\alpha_i = 1$ for all

$i=1,\ldots,n$, then we have

$$P(x_1 \geqslant t_1,\ldots,x_n \geqslant t_n) = W^n g(t_1+\cdots+t_n)$$
$$= \int_{\sum t_i}^{\infty} \frac{(r-\sum t_i)^{n-1}}{(n-1)!} g(r)\mathrm{d}r, \tag{6.77}$$

which is the same as (5.12). Hopefully, more light can be shed on the survival function after a thorough study of the extended Weyl operator (6.75). This approach is at present being investigated by a co-author of this book, K.W. Ng.

Example 6.17 (Inverted Dirichlet distribution)
Let $\mathbf{x}$ be distributed as $ID_n(\alpha_1,\ldots,\alpha_n;\beta,\lambda)$ so that its density generator is (cf. Example 6.3)

$$g(r) = \frac{\Gamma(\alpha+\beta)}{\Gamma(\alpha)\Gamma(\beta)} \frac{\lambda^\beta}{(\lambda+r)^{\alpha+\beta}}, \qquad \alpha = \sum_{i=1}^n \alpha_i$$

Making use of the inverted Dirichlet density (6.14), we have

$$W^\delta g(t) = \frac{\Gamma(\alpha+\beta-\delta)}{\Gamma(\beta)} \frac{\lambda^\beta}{(\lambda+t)^{\alpha+\beta-\delta}} \tag{6.78}$$

where $0 \leqslant \delta \leqslant \alpha+\beta$. If the Dirichlet parameters $\alpha_1,\ldots,\alpha_n$ are all integers, Theorem 6.20(d) implies that

$$S(t_1,\ldots,t_n) = \sum_{k_1=0}^{\alpha_1-1} \cdots \sum_{k_n=0}^{\alpha_n-1} \frac{t_1^{k_1}\ldots t_n^{k_n}}{k_1!\ldots k_n!} W^{\alpha-\sum k_i} g\left(\sum_{i=1}^n t_i\right). \tag{6.79}$$

Substituting (6.78) into (6.79), we get

$$S(t_1,\ldots,t_n) = \frac{1}{\Gamma(\beta)}\left(\frac{\lambda}{\lambda+\sum t_i}\right)^\beta \sum_{k_1=0}^{\alpha_1-1} \cdots \sum_{k_n=0}^{\alpha_n-1} \frac{t_1^{k_1}\ldots t_n^{k_n}}{k_1!\ldots k_n!} \frac{\Gamma(\beta+\sum k_i)}{(\lambda+\sum t_i)^{\sum k_i}}. \tag{6.80}$$

In particular, the reliability of a series system whose components have an inverted Dirichlet distribution is given by

$$P(x_i \geqslant t \quad \text{for } 1 \leqslant i \leqslant n)$$
$$= \frac{1}{\Gamma(\beta)}\left(\frac{\lambda}{\lambda+nt}\right)^\beta \sum_{k_1=0}^{\alpha_n-1} \cdots \sum_{k_n=0}^{\alpha_n-1} \frac{\Gamma(\beta+\sum k_i)}{k_1!\ldots k_n!}\left(\frac{t}{\lambda+nt}\right)^{\sum k_i}. \tag{6.81}$$

When $\alpha_i = 1$, $i=1,\ldots,n$, (6.80) simplifies to

$$S(t_1,\ldots,t_n) = \left(\frac{\lambda}{\lambda+\sum t_i}\right)^\beta. \tag{6.82}$$

Example 6.18 (*Predicting the reliability of a series system*)
If the predictive distribution of $(x_{m+1},\ldots,x_n)$ given $(x_1,\ldots,x_m)$ in Example 6.15 has integer parameters α_i, $i=m+1,\ldots,n$, we have

$$P\left(x_j \geqslant t\sum_{i=1}^{m} x_i, m<j\leqslant n\right) = \frac{(1+(n-m)t)^{-\alpha_1^*}}{\Gamma(\alpha_1^*)}\sum_{k_{m+1}=0}^{\alpha_{m+1}-1}\cdots\sum_{k_n=0}^{\alpha_n-1}\frac{\Gamma(\alpha_1^*+\sum k_i)}{k_{m+1}!\ldots k_n!}\left(\frac{t}{1+(n-m)t}\right)^{\sum k_i}, \tag{6.83}$$

where $\alpha_1^*=\sum_{i=1}^m \alpha_i$. This is useful for applications where m items of a batch of n are subject to destructive testing to get information about the items' lifetime and then the remaining $n-m$ items will be used as components in a series system. In particular, when $\alpha_i=1$ for $i=1,\ldots,n$, we have the following result obtained by Lawless (1972) who assumes independent exponential distributions for the components:

$$P\left(x_j \geqslant t\sum_{j=1}^{m} x_i, \quad m<j\leqslant n\right) = (1+(n-m)t)^{-m}. \tag{6.84}$$

When α_i are not integers, a lower bound for the predicted reliability (6.83) can be found using an inequality to be presented in the next section (cf. Example 6.20).

6.8 Inequalities and applications

To fill the computational gap, we need some bounds for the probabilities of interest. One naturally thinks of the inequalities of Bonferroni type. Here we state without proof the results by Gallot (1966) and Kounias (1968). See Tong (1980) for other Bonferroni-type inequalities, and Hochberg and Tamhane (1987, p. 364) for Hunter–Worsley's improvement over the Kounias inequality.

Theorem 6.21
Let $(x_1,\ldots,x_n)$ be a random vector and let $q_i=P(B_i)$ and $q_{ij}=P(B_i\cap B_j)$, where the event B_i, $i=1,\ldots,n$, is concerned with x_i alone. Then the following inequalities are valid:

(a) $$P\left(\bigcup_{i=1}^{n} B_i\right) \geqslant \mathbf{q}'\mathbf{Q}^{-}\mathbf{q} \geqslant \sum_{i=1}^{n} q_i \Bigg/ \sum_{i=1}^{n}\sum_{j=1}^{n} q_{ij},$$

where $\mathbf{q}=(q_i)$, $\mathbf{Q}=(q_{ij})$, and $\mathbf{Q}^-$ is any generalized inverse of $\mathbf{Q}$.

(b) $P\left(\bigcup_{i=1}^{n} B_i\right) \geqslant \sum_{i=1}^{n} q_i - \sum_{i=1}^{n}\sum_{j=1}^{n} q_{ij}.$

(c) $P\left(\bigcup_{i=1}^{n} B_i\right) \leqslant \sum_{i=1}^{n} q_i - \max_{1\leqslant i\leqslant n} \sum_{j\neq i} q_{ij} \leqslant \sum_{i=1}^{n} q_i.$

For an intersection of events, we may use $P(\cap A_i) = 1 - P(\cup B_i)$, where A_i is the complement of B_i. Note that part (c) of the above theorem implies the Bonferroni inequality

$$P\left(\bigcap_{i=1}^{n} A_i\right) \geqslant 1 - \sum_{i=1}^{n} (1 - P(A_i)). \tag{6.85}$$

The drawback of these inequalities, except (6.85) which is generally over-conservative when n is moderate to large, is the computation involved with the bivariate distributions. Karlin and Rinott (1980a, 1980b) obtain some inequalities, in terms of univariate distributions only, for multivariate distributions which are called **multivariate total positive of order 2** (MTP_2) and **strongly multivariate reverse rule of order 2** (S-MRR_2). For the Liouville distributions, Gupta and Richards (1987) give a necessary and sufficient condition for the MTP_2 condition and a sufficient condition for the S-MRR_2 condition in terms of the logarithmic convexity of the density generator. These important results make it possible that the inequalities of Karlin and Rinott can be applied to the Liouville distributions. It should be noted that the MTP_2 condition is equivalent to the FKG condition, named after the paper by Fortuin, Kastelyn, and Ginibre (1971), which contains some of the results by Karlin and Rinott. See also the papers by Kemperman (1977) and Perlman and Olkin (1980). In the theorems that follow, we shall present without proof the inequalities obtained by Gupta and Richards (1987).

Theorem 6.22
Let $\mathbf{x}$ be distributed as $L_n(g; \alpha_1, \ldots, \alpha_n)$, where the density generator $g(\cdot)$ is a logarithmically convex function over $(0, \infty)$. Then the following statements are true:

(a) The density generator of a subvector of $\mathbf{x}$ is also logarithmically convex.
(b) If h is a function which is nondecreasing in each of its arguments,

the multiple regression function

$$E(h(x_1, \ldots, x_r | x_{r+1}, \ldots, x_n))$$

is nondecreasing in each of $x_{r+1}, \ldots, x_n$.

(c) If h_1 and h_2 are both nondecreasing or both nonincreasing functions in each of its arguments, then

$$\mathrm{Cov}(h_1(\mathbf{x}), \quad h_2(\mathbf{x})) \geqslant 0. \tag{6.86}$$

(d) Furthermore, if $h_1, \ldots, h_m$ are all nondecreasing or all nonincreasing in each of its arguments, then

$$E\left(\prod_{i=1}^{m} h_i(\mathbf{x})\right) \geqslant \prod_{i=1}^{m} E(h_i(\mathbf{x})). \tag{6.87}$$

In particular,

$$P(x_1 \geqslant t_1, \ldots, x_n \geqslant t_n) \geqslant \prod_{i=1}^{n} P(x_i \geqslant t_i), \tag{6.88}$$

$$P(x_1 \leqslant t_1, \ldots, x_n \leqslant t_n) \geqslant \prod_{i=1}^{n} P(x_i \leqslant t_i), \tag{6.89}$$

Note that properties (c) and (d) are equivalent, although (d) looks more general than (c). Choosing $h_i(\mathbf{x})$ as the indicator function of the set $\{\mathbf{x}: x_i \geqslant t \text{ for all } i\}$ or the set $\{\mathbf{x}: x_i \leqslant t \text{ for all } i\}$, we get (6.88) or (6.89) by means of (6.87).

Example 6.19 (*Logarithmically convex generators*)
Since the second derivative of the logarithm of (6.15) is positive for all $r > 0$ if an only if $\alpha \leqslant \alpha^*$, we see that the density generator of an $IBL_n(\alpha_1, \ldots, \alpha_n;\ \alpha, \beta, \lambda)$ is logarithmically convex if and only if $\alpha \leqslant \sum_{i=1}^n \alpha_i$. Next, we consider a Weibull $\mathrm{We}(\beta, \lambda)$ generating density

$$f(r) = \beta\lambda^\beta r^{\beta-1} \exp(-\lambda^\beta r^\beta), \qquad \beta > 0, \lambda > 0. \tag{6.90}$$

Its corresponding density generator,

$$g(r) = \Gamma(\alpha)\beta\lambda^\beta r^{\beta-\alpha} \exp(-\lambda^\beta r^\beta), \qquad \alpha = \sum_{i=1}^{n} \alpha_i, \tag{6.91}$$

is logarithmically convex if and only if

$$(\beta - \alpha) + \beta(\beta - 1)\lambda^\beta r^\beta < 0$$

for all r, which in turn is equivalent to the condition that $\beta < \min(1, \alpha)$.

Example 6.20 (Continuation of Example 6.15)
Since the marginal distribution of $x_j/\sum_{i=1}^m x_i$, $j = m+1, \ldots, n$ is $ID(\alpha_j; \sum_{i=1}^m \alpha_i, 1)$, we can use (6.88) to get a lower bound of the survival function

$$P\left(x_j \geqslant a_j \sum_{i=1}^m x_i, m < j \leqslant n\right) \geqslant \prod_{j=m+1}^n P\left(w_j \geqslant \frac{a_j}{1-a_j}\right), \tag{6.92}$$

where w_j is distributed as $\mathrm{Be}(\alpha_j, \sum_{i=1}^m \alpha_i)$, $j = m+1, \ldots, n$. In particular, the reliability (6.83) has a lower bound,

$$P\left(x_j \geqslant t \sum_{i=1}^m x_i, m < j \leqslant n\right) \geqslant \prod_{i=1}^n P\left(w_j \geqslant \frac{t}{1-t}\right). \tag{6.93}$$

Since the right-hand side is a monotone function, we can find t for any given value of the lower bound by plotting and/or an iterative search program. Analogously, the inequality (6.89) gives a lower bound for the distribution function,

$$P\left(x_j \leqslant a_j \sum_{i=1}^m x_i, m < j \leqslant n\right) \geqslant \prod_{i=1}^n P\left(w_j \leqslant \frac{a_j}{1-a_j}\right). \tag{6.94}$$

Example 6.21 (Multiple Type I risk in ANOVA)
It is a common procedure in the analysis of variance for orthogonal designs that we test several hypotheses based on their corresponding F statistics having the same denominator, the error sum of squares. The multiple Type I risk of the tests (that is, the probability of claiming at least one significance incorrectly) is actually higher than the significance level we set for each test. Note that the tests are not mutually independent because the same error sum of squares is used as denominator. When the hypotheses considered are true, all the sums of squares, denoted as x_i, $i = 1, \ldots, k, k+1$ are independently chi-squared, i.e. independent gamma distributions with a common scale parameter. Let $y_i = x_i/x_{k+1}$, $i = 1, \ldots, k$, then $(y_1, \ldots, y_k)$ has an inverted Dirichlet distribution (cf. Example 6.15) possessing a logarithmically convex density generator. Using (6.89), we have

$$1 - P(y_1 \leqslant a_1, \ldots, y_k \leqslant a_k) \leqslant 1 - \prod_{i=1}^k (1 - P(y_i > a_i)). \tag{6.95}$$

Evidently, this inequality holds for the F statistics, which are scalar multiples of the y_i. Let α_i be the significance level for the ith test,

$i = 1, \ldots, k$. The above inequality becomes the following inequality of Kimball (1951),

$$\text{Multiple Type I risk} \leqslant 1 - \prod_{i=1}^{k} (1 - \alpha_i). \tag{6.96}$$

In general, this upper bound is smaller (thus better) than the omnibus Benferroni upper bound. Conversely, if a maximum α is set on the multiple Type I risk, we may choose for all tests a common significance level α_0 satisfying

$$\alpha_0 = 1 - (1 - \alpha)^{1/k}. \tag{6.97}$$

The following result of Gupta and Richards concerns logarithmically concave density generators.

Theorem 6.23

Let $\mathbf{x}$ be distributed as $L_n(g; a_1, \ldots, a_n)$, where $a_i \geqslant 1$ and the density generator g is a monotone and logarithmically concave function defined over a finite interval $(0, c)$. If $h_1, \ldots, h_n$ are all nondecreasing functions then for $m = 1, 2, \ldots, n$,

$$E\left(\prod_{i=1}^{n} h_i(x_i)\right) \leqslant E\left(\prod_{i=1}^{m} h_i(x_i)\right) E\left(\prod_{i=m+1}^{n} h_i(x_i)\right) \leqslant \prod_{i=1}^{n} h_i(x_i), \tag{6.98}$$

provided the expectations exist. In particular,

$$P(x_1 \geqslant t_1, \ldots, x_n \geqslant t_n) \leqslant \prod_{i=1}^{n} P(x_i \geqslant t_i). \tag{6.99}$$

Example 6.22 (Beta Liouville distribution)

The first and second derivatives of $\log g(r)$ for the density generator (6.12) are respectively

$$(\alpha - \alpha^*) r^{-1} - (\beta - 1)(1 - r)^{-1}$$

and

$$-(\alpha - \alpha^*) r^{-2} + (\beta - 1)(1 - r)^{-2}.$$

Therefore, $g(r)$ is both logarithmically concave and monotone if and only if $\alpha \geqslant \sum_{i=1}^{n} \alpha_i$ and $\beta < 1$, or $\alpha < \sum_{i=1}^{n} \alpha_i$ and $\beta \leqslant 1$. Under these conditions, we may give upper bounds of the survival function of the smallest component of a Dirichlet vector (cf. Example 6.16) using (6.99).

Problems

6.1 Find the means, variances and covariances of the following multivariate distributions:

(a) beta Liouville $BL_n(\alpha_1, \ldots, \alpha_n; \alpha, \beta)$;
(b) inverted beta Liouville $IBL_n(\alpha_1, \ldots, \alpha_n; \alpha, \beta, \lambda)$;
(c) gamma Liouville $GL_n(\alpha_1, \ldots, \alpha_n; \alpha, \lambda)$;
(d) exponential-gamma Liouville $EGL_n(\alpha_1, \ldots, \alpha_n; \alpha, \lambda)$.
Hint: Apply Theorem 6.3.

6.2 Let an $n \times 1$ random vector $\mathbf{y}$ be partitioned into $n_i \times 1$ subvectors $\mathbf{y}_i$, $i = 1, \ldots, m$, $n = \sum n_i$. Verify that if $\mathbf{y}$ has a standard multivariate $t(\lambda)$ distribution with density function

$$p(\mathbf{y}) = c(n, \lambda)(1 + \lambda^{-1}\mathbf{y})^{-(n+\lambda)/2},$$

the joint distribution of the m sums of squares $(\mathbf{y}_1'\mathbf{y}_1, \ldots, \mathbf{y}_m'\mathbf{y}_m)$ has an inverted Dirichlet $ID_m(n_1/2, \ldots, n_m/2; \lambda, \lambda)$; in particular, $\lambda^{-1}\sum_{j=1}^n y_j^2 \sim ID_n(\frac{1}{2}, \ldots, \frac{1}{2}; \lambda, 1)$.
Hint: Make use of (3.25), Examples 6.6 and 6.3.
The special case is a result of Mihoc (1987).

6.3 Verify Theorem 6.4(ii).

6.4 Show that if x has a $\mathrm{Be}(\alpha, \beta)$ distribution, then $E(x^{it})$ is nonzero for all real t.

6.5 Derive (6.28) using the stochastic representation (6.33).
Hint: Put $k = 2$ in (6.33), then $w_1 \sim \mathrm{Be}(\alpha_1^*, \alpha_2^*)$. Show that the mixture rw_1 has density (6.28).

6.6 Verify (6.36), (6.37) and (6.39) in detail.

6.7 Verify the conditional moments (6.50) to (6.52).
Hint: Apply the integral

$$\int x^m(a+bx)^c \,\mathrm{d}x = \{b(m+c+1)\}^{-1} \times \left\{x^m(a+bx)^{c+1} - ma\int x^{m-1}(a+bx)^c \,\mathrm{d}x\right\}.$$

6.8 Verify the conditional moments (6.54) to (6.56).
Hint: Apply the following integrals

$$\int \frac{x\,\mathrm{d}x}{(a+bx)^c} = \frac{1}{b^2}\left\{\frac{-1}{(c-2)(a+bx)^{c-2}} + \frac{a}{(c-1)(a+bx)^{c-1}}\right\}, \quad c \neq 1, 2;$$

$$\int \frac{x^2\,dx}{(a+bx)^c} = \frac{1}{b^3}\left\{\frac{-1}{(c-3)(a+bx)^{c-3}} + \frac{2a}{(c-2)(a+bx)^{c-2}} - \frac{a^2}{(c-1)(a+bx)^{c-1}}\right\}, \qquad c \neq 1, 2, 3.$$

6.9 Solve the Cauchy multiplicative equation (6.62).
Hint: Make use of the solution $g(x) = cx$ for the Cauchy functional equation

$$g(t_1 + t_2) = g(t_1) + g(t_2),$$

where $g(\cdot)$ is any continuous function defined over an open interval.

6.10 Prove Theorem 6.14 in detail by mimicking the proof of Theorem 6.13.

6.11 Show that if $\mathbf{x}$ has a Liouville distribution with Dirichlet parameters $(\alpha_1, \ldots, \alpha_n)$, the statistic $t = x_n / \sum_{i=1}^{n-1} x_i$ has an inverted beta distribution $IB(\alpha; \sum_{i=1}^{n-1} \alpha_i, 1)$ (see Example 6.3); so that $W = t/(1+t)$ has $\mathrm{Be}(\alpha_n, \sum_{i=1}^{n-1} \alpha_i)$ distribution.

6.12 Verify Theorem 6.20(d).

CHAPTER 7

α-Symmetric distributions

The α-symmetric multivariate distribution has been introduced in Section 1.1, as a natural generalization of an i.i.d. sample of symmetric stable law (1.8). In fact, the spherical distribution is a special case of the α-symmetric distribution with $\alpha = 2$. The progress of α-symmetric distributions for other cases of α has been relatively slow, perhaps due to the fact that the analysis with characteristic functions heavily involves calculus of complex variables which is not a favourite tool among statisticians. This may change, however, especially when statisticians realize that distributions encountered in sciences and social sciences need not possess nice explicit p.d.f.s nor finite moments, and that it is still possible to do statistical analysis with such distributions. The excellent exposition on one-dimensional stable distributions by Zolotarev (1985) may well attract the attention of statisticians in this direction. In this last chapter, we shall present the general properties of α-symmetric distributions and consider some special cases of α and n. We emphasize that the problems listed at the end will form an integral part of this chapter and will be referred to from time to time. The results presented here are mainly taken from Cambanis, Keener and Simons (1983) and Kuritsyn and Shestakov (1984). Some unpublished results of K.W. Ng are also incorporated in the presentation.

7.1 α-Symmetric distributions

There is more than one natural way to generalize a univariate symmetric stable law which has characteristic function (cf. Zolotarev, 1985)

$$\exp(-\lambda|t_1|^{\alpha}), \qquad 0 < \alpha \leqslant 2. \tag{7.1}$$

One may replace the absolute value by the ℓ_2-norm of $\mathbf{t}$, obtaining what is called a **spherically symmetric stable law**. This is a special case of the spherical distributions, as mentioned in Chapter 3. A detailed discussion on these laws is given by Zolotarev (1981) and more examples occurring in applications can be found in Zolotarev (1985, Ch. 1). Alternatively, one may view (7.1) as a function of an ℓ_α-norm of **t**, but with dimension $n = 1$. This way of generalization leads to a characteristic function in terms of an ℓ_α-norm,

$$\psi(\|\mathbf{t}\|_\alpha), \ \|\mathbf{t}\|_\alpha = (\sum |t_i|^\alpha)^{1/\alpha}, \tag{7.2}$$

where $\psi(\cdot)$ is a function from R_+ to R. Notice that if **x** has a c.f. given by (7.2), $\psi(|t|)$ is the common c.f. of the components x_i $(i = 1, \ldots, n)$. As a third approach, one may take an i.i.d. sample of the symmetric stable law (7.1), obtaining an n-dimensional c.f.

$$\exp(-\lambda \sum |t_i|^\alpha) = \exp(-\lambda \|\mathbf{t}\|_\alpha^\alpha), \tag{7.3}$$

which is also in the form of (7.2) with $\psi(u) = \exp(-\lambda u^\alpha)$.

Definition 7.1
An n-dimensional distribution whose c.f. is in the form of (7.2) is called **α-symmetric with c.f. generator** $\psi(\cdot)$. If **x** has an α-symmetric distribution, this fact shall be expressed symbolically as $\mathbf{x} \sim S_n(\alpha, \psi)$. For a fixed n, the class of all functions $\psi(\cdot)$ from R_+ to R such that $\psi(\|\mathbf{t}\|_\alpha)$ is a c.f. for some n-dimensional distribution will be denoted by $\Psi_n(\alpha)$. The number $\alpha > 0$ is called the **index**, or **norm index**.

In the original definition of Cambanis, Keener and Simons (1983), the c.f. of an α-symmetric distribution is expressed as

$$\phi(\sum |t_i|^\alpha), \tag{7.4}$$

where $\phi(\cdot)$ is called a 'primitive', so that the c.f. generator ψ is related to ϕ as follows:

$$\psi(t) = \phi(t^\alpha), \qquad \psi(t^{1/\alpha}) = \phi(t). \tag{7.5}$$

The definition here emphasizes the relation to the ℓ_α-norm and conveniently accommodates the class $\Psi_n(\infty)$ which corresponds to the maximum norm $\|\mathbf{t}\|_\infty = \max\{|t_1|, \ldots, |t_n|\}$. Note that since we do not make use of the triangle inequality of a norm, we still call it a norm when $0 < \alpha < 1$, for convenience. The following theorem confirms the assertion in Section 1.1 that an α-symmetric distribution is a special n-dimensional version of a univariate distribution.

Theorem 7.1

The following statements are true:

(i) $\mathbf{x} \sim S_n(\alpha, \psi)$ if and only if $\mathbf{a}'\mathbf{x} \sim S_1(\alpha, \psi)$ for all $\mathbf{a}$ such that $\|\mathbf{a}\|_\alpha = 1$. Consequently, an α-symmetric distribution is uniquely determined by any of its marginal distributions.

(ii) $\mathbf{x}$ is α-symmetric if and only if $\mathbf{a}'\mathbf{x} \overset{d}{=} \|\mathbf{a}\|_\alpha x_1$ for all $\mathbf{a}$.

PROOF

(i) If $\mathbf{x} \sim S_n(\alpha, \psi)$, the c.f. of $\mathbf{a}'\mathbf{x}$ is

$$E(e^{it\mathbf{a}'\mathbf{x}}) = \psi(\|t\mathbf{a}\|_\alpha) = \psi(|t|\,\|\mathbf{a}\|_\alpha) = \psi(|t|).$$

Conversely, let $\mathbf{a} = \|\mathbf{t}\|_\alpha^{-1}\mathbf{t}$, we then compute the c.f. of $\mathbf{x}$:

$$E(e^{i\mathbf{t}'\mathbf{x}}) = E(e^{i\|\mathbf{t}\|_\alpha \mathbf{a}'\mathbf{x}}) = \psi(\|\mathbf{t}\|_\alpha).$$

The proof of (ii) is similar and is left to the reader. □

Corollary 1

Let $\mathbf{y}$ be a random vector independent of $\mathbf{x} \sim S_n(\alpha, \psi)$, both having an equal dimension. The vector $\|\mathbf{y}\|_\alpha^{-1}\mathbf{y}'\mathbf{x} \overset{d}{=} x_1$, which is independent of $\mathbf{y}$.

Corollary 2

If a marginal distribution of an α-symmetric vector $\mathbf{x}$ is a symmetric stable law (7.1), all components of $\mathbf{x}$ have i.i.d. symmetric stable distributions.

PROOF Since the c.f. generator is given as $\psi(u) = \exp(-\lambda u^\alpha)$, the c.f. of $\mathbf{x}$ is $\psi(\|\mathbf{t}\|_\alpha) = \exp(-\lambda \sum |t_i|^\alpha)$, showing an i.i.d. sample of a symmetric stable law. □

Another role that the symmetric stable law plays in the α-symmetric family is exhibited by the following theorem.

Theorem 7.2

If an α-symmetric vector $\mathbf{x}$ contains two independent subvectors, all the components have i.i.d. symmetric stable distribution (7.1).

PROOF Let x_i be any component in one of the subvectors, and x_j be any one in the other. Then $(x_i, x_j) \sim S_2(\alpha, \psi)$ and at the same time they are independent. This implies that

$$\psi((t_1^\alpha + t_2^\alpha)^{1/\alpha}) = \psi(t_1)\psi(t_2), \qquad t_1 > 0, \quad t_2 > 0. \tag{7.6}$$

Let $g(u) = \psi(u^{1/\alpha})$, and $u_i = t_i^\alpha$. The above functional equation is represented as

$$g(u_1 + u_2) = g(u_1)g(u_2), \qquad u_1 > 0, \quad u_2 > 0. \tag{7.7}$$

The nonzero continuous solution for this Cauchy functional equation is $g(u) = e^{cu}$, so that $\psi(t) = e^{ct^\alpha}$ for $t > 0$. Since $\psi(|t|)$ is a c.f., $|\psi(t)| < 1$ for $t > 0$, hence $c = -\lambda$ for some $\lambda > 0$. This completes the proof. □

As we have seen in Chapter 2, the spherical distribution, which is α-symmetric with $\alpha = 2$, can be represented as an independent product of a nonnegative generating variate and a spherical vector base. Can an analogous decomposition $\mathbf{x} = r\mathbf{y}$ exist for other cases of α-symmetric distributions? In an important paper containing many interesting results, Cambanis *et al.* (1983) obtained the stochastic decomposition for the case $\alpha = 1$. Some discussion for the case $\alpha = 1/2$ was also given. The existence of such a stochastic decomposition for a nonparametric family of distributions is very useful in that it identifies the scale-invariant statistics as the only statistics that have fixed distributions.

Theorem 7.3

Let a family of multivariate distributions for $\mathbf{x}$ have the stochastic representation $\mathbf{x} \stackrel{d}{=} r\mathbf{y}$, where $r > 0$ and $\mathbf{y}$ are independent, and $\mathbf{y}$ has a fixed distribution for all members in the family. The distribution of a k-dimensional statistic $\mathbf{t}(\mathbf{x})$ is fixed for all members in the family if and only if $\mathbf{t}(c\mathbf{x}) \stackrel{d}{=} \mathbf{t}(\mathbf{x})$ for all $c > 0$.

The proof of this theorem is almost the same as that for Theorem 6.19 and will be omitted.

The following theorem presents a criterion for establishing a derived membership in an α-symmetric family, which will play an important role in the next section.

Theorem 7.4

Let $\mathbf{x} \stackrel{d}{=} r\mathbf{y}$, where $\mathbf{y}$ is independent of $r > 0$. If $\mathbf{y}$ is α-symmetric, so is $\mathbf{x}$. Conversely, if $\mathbf{x}$ is α-symmetric and if $E(r^{it})$ is nonzero for almost all t, then $\mathbf{y}$ is α-symmetric.

PROOF The first part is straightforward. Conversely, assume $\mathbf{x}$ is α-symmetric. Then for all $\mathbf{a}$ such that $\|\mathbf{a}\|_\alpha = 1$, we have $\mathbf{a}'\mathbf{x} \stackrel{d}{=} x_1$, so that $r\mathbf{a}'\mathbf{y} = ry_1$. If $E(r^{it})$ is nonzero almost everywhere, we can apply Lemma 1.2 (ii) to conclude that $\mathbf{a}'\mathbf{y} \stackrel{d}{=} y_1$. The proof is completed by applying Theorem 7.1. □

The last three properties, observed by K.W. Ng, seem to be new.

7.2 Decomposition of 1-symmetric distributions

In this section we shall present the decomposition theorem of Cambanis, Keener and Simons for the class $\Psi_n(1)$ of 1-symmetric distributions. However, we do not follow exactly their route. By rearranging the flow of arguments and filling up some gaps, we obtain a derivation which seems more economical. The following pivotal result is due to Cambanis *et al.* (1983, p. 225).

Lemma 7.1
Let $\mathbf{x} \sim S_n(1, \psi(\cdot))$, where $\psi(\cdot)$ satisfies the condition

$$\int_0^\infty u^{n-1}|\psi(u)|\,\mathrm{d}u < \infty. \tag{7.8}$$

The p.d.f. of $\mathbf{x}$ is given by

$$p(\mathbf{x}) = \int_0^\infty r^{-n} p_0(r^{-1}\mathbf{x}) f(r)\,\mathrm{d}r, \tag{7.9}$$

where

$$p_0(\mathbf{y}) = \frac{\Gamma^2(n/2)}{(n-2)!\,\pi^n} \sum_{k=1}^n \left\{ \frac{(y_k^2 - 1)_+^{n-2}}{\prod_{j \neq k} (y_k^2 - y_j^2)} \right\}, \qquad |y_k| \neq |y_j|, \quad k \neq j, \tag{7.10}$$

$$f(r) = \frac{2}{\Gamma^2(n/2)} r^{n-1} B_n^{(n-1)}(r^2), \qquad r > 0, \tag{7.11}$$

and

$$B_n(t) = \begin{cases} (-1)^{(n-2)/2} t^{(n-1)/2} \displaystyle\int_0^\infty \sin(u\sqrt{t})\psi(u)\,\mathrm{d}u, & n \text{ even}, \\ (-1)^{(n-1)/2} t^{(n-1)/2} \displaystyle\int_0^\infty \cos(u\sqrt{t})\psi(u)\,\mathrm{d}u, & n \text{ odd}. \end{cases} \tag{7.12}$$

PROOF It follows from (7.8) that the c.f. of $\mathbf{x}$ is absolutely integrable in R^n:

$$\int_{-\infty}^{\infty}\cdots\int_{\infty}^{\infty}|\psi(|t_1|+\cdots+|t_n|)|\mathrm{d}t_1\ldots\mathrm{d}t_n = \frac{2^n}{(n-1)!}\int_0^\infty u^{n-1}|\psi(u)|\mathrm{d}u<\infty.$$

Hence by the inversion formula for p.d.f., we have

$$\begin{aligned} p(\mathbf{x}) &= (2\pi)^{-n}\int_{-\infty}^{\infty}\cdots\int_{-\infty}^{\infty} e^{-i(\sum t_i x_i)}\psi(\textstyle\sum|t_i|)\,\mathrm{d}t_1\ldots\mathrm{d}t_n \\ &= \pi^{-n}\int_0^\infty\cdots\int_0^\infty \psi\left(\sum_{i=1}^n t_i\right)\prod_{i=1}^n \cos(t_i x_i)\,\mathrm{d}t_i, \end{aligned}$$

where in the last step we have used the symmetry of the integrand. By using the 'Even' operation on functions:

$$\operatorname{Even} h(x_1,\ldots,x_n) = 2^{-n}\sum h(\pm x_1,\cdots\pm x_n),$$

where the sum is over all 2^n possible choices of signs, we have

$$p(\mathbf{x}) = \pi^{-n}\operatorname{Even}\int_0^\infty\cdots\int_0^\infty \psi\left(\sum_{i=1}^n t_i\right)\exp(i\textstyle\sum t_i x_i)\,\mathrm{d}t_1\ldots\mathrm{d}t_n. \quad (7.13)$$

Using the integral given in Problem 7.3, we obtain

$$\pi^n p(\mathbf{x}) = \operatorname{Even}\int_0^\infty \psi(u) i^{-(n-1)}\sum_{k=1}^n \frac{\exp(iux_k)}{\prod_{j\neq k}(x_k - x_j)}\mathrm{d}u.$$

Since the Even operator is effectively a summation over functions of $x_1,\ldots,x_n$, it operates directly on the last factor of the integrand,

$$\operatorname{Even}\sum_{k=1}^n \frac{\exp(iux_k)}{\prod_{j\neq k}(x_k-x_j)} = \begin{cases} i\displaystyle\sum_{k=1}^n \frac{x_k^{n-1}\sin(ux_k)}{\prod_{j\neq k}(x_k^2-x_j^2)}, & n \text{ even},\\[2ex] \displaystyle\sum_{k=1}^n \frac{x_k^{n-1}\cos(ux_k)}{\prod_{j\neq k}(x_k^2-x_j^2)}, & n \text{ odd}. \end{cases}$$

Therefore, we can write

$$\pi^n p(\mathbf{x}) = \sum_{k=1}^n \frac{B_n(x_k^2)}{\prod_{j\neq k}(x_k^2-x_j^2)}, \quad (7.14)$$

where the function $B_n(t)$ is given in (7.12). It follows from (7.8) that the following Taylor expansion is valid:

$$B_n(t) = \sum_{k=0}^{n-2} \frac{t^k}{k!} B_n^{(k)}(0+) + \frac{1}{(n-2)!} \int_0^t (t-u)^{n-2} B_n^{n-1}(u)\, du. \qquad (7.15)$$

Note that the right-hand side of (7.14) is the $(n-1)$th divided difference of the function $B_n(\cdot)$ at the points $x_1^2, \ldots, x_n^2$ and that the first term of (7.15) is a polynomial of degree $(n-2)$ whose $(n-1)$th divided difference is zero. Therefore, substituting (7.15) into (7.14), we obtain

$$\begin{aligned}\pi^n p(\mathbf{x}) &= \frac{1}{(n-2)!} \int_0^\infty \sum_{k=1}^n \frac{(x_k - u)_+^{n-2}}{\prod_{j \neq k} (x_k^2 - x_j^2)} B_n^{(n-1)}(u)\, du \\ &= \frac{1}{(n-2)!} \int_0^\infty \sum_{k=1}^n \frac{(x_k^2/r^2 - 1)_+^{n-2}}{\prod_{j \neq k} (x_k^2/r^2 - x_j^2/r^2)} B_n^{(n-1)}(r^2) \frac{2}{r}\, dr.\end{aligned}$$

This is exactly the integral representation (7.9), hence the proof is completed. □

Note that, in Lemma 7.1, the function $f(r)$ depends on $\psi(\cdot)$, but $p_0(\mathbf{x})$ does not. If we can show that both f and p_0 are p.d.f.s, then the independent decomposition $\mathbf{x} = r\mathbf{y}$ follows, with r having density $f(r)$ and $\mathbf{y}$ having density $p_0(\mathbf{y})$. Furthermore, since $p(\cdot)$ is a p.d.f., it suffices to show that $f(\cdot)$ is a p.d.f. For $\alpha = 1$, the stable symmetric distribution reduces to the standard Cauchy distribution having c.f. $\exp(-|t|)$. Being a standard Student's $t(1)$ variable, the standard Cauchy variable $x \overset{d}{=} z_1/|z_2|$, where z_1 and z_2 are two independent standard normal variables. So for an i.i.d. sample $(x_1, \ldots, x_n)$ of standard Cauchy distributions, we have

$$(x_1, \ldots, x_n) \overset{d}{=} \left(\frac{z_{11}}{|z_{21}|}, \ldots, \frac{z_{1n}}{|z_{2n}|} \right),$$

where z_{ij} are i.i.d. $N(0,1)$. Applying the decomposition for spherical distributions (Chapter 2) to $(z_{i1}, \ldots, z_{in})$ separately, $i = 1, 2$, we obtain

$$(x_1, \ldots, x_n) \overset{d}{=} \frac{r_1}{r_2} \left(\frac{u_1}{|v_1|}, \ldots, \frac{u_n}{|v_n|} \right), \qquad (7.16)$$

where $r_1, r_2, (u_1, \ldots, u_n)$ and $(v_1, \ldots, v_n)$ are independent, both r_1^2 and r_2^2 have $\chi^2(n)$ distribution, and both vectors have uniform distribution on the surface of a unit sphere in R^n. Note that (cf. Section 3.1.3) $(v_1^2,\ldots,v_n^2) \sim D_n(\frac{1}{2},\ldots,\frac{1}{2})$, so that (7.16) becomes

$$(x_1,\ldots,x_n) \stackrel{d}{=} r\left(\frac{u_1}{\sqrt{w_1}},\ldots,\frac{u_n}{\sqrt{w_n}}\right) = r\mathbf{y}', \tag{7.17}$$

where $r^2 \sim F(n,n)$, $\mathbf{u}$ is uniform on unit sphere S_n, $(w_1,\ldots,w_n)$ is $D_n(\frac{1}{2},\ldots,\frac{1}{2})$, and they are independent. This suggests that we apply Lemma 7.1 to $\mathbf{x}$ and check whether $f(r)$ coincides with the distribution for the square root of an $F(n,n)$ variable. If it does, we can then conclude that the functions $f(\cdot)$ and $p_0(\cdot)$ in Lemma 7.1 are both p.d.f.s and that $p_0(\mathbf{y})$ is indeed the p.d.f. of vector base $\mathbf{y}$ in (7.17).

Lemma 7.2

The functions $p_0(\cdot)$ and $f(\cdot)$ in Lemma 7.1 are both p.d.f.s. Furthermore, $p_0(\cdot)$ is the p.d.f. of

$$\mathbf{y}' = \left(\frac{u_1}{\sqrt{w_1}},\ldots,\frac{u_n}{\sqrt{w_n}}\right), \tag{7.18}$$

where $(w_1,\ldots,w_n) \sim D_n(\frac{1}{2},\ldots,\frac{1}{2})$ is independent of $(u_1,\ldots,u_n)$ which has a uniform distribution on the unit sphere in R^n.

PROOF Let $\mathbf{x}$ have c.f. generator $\psi(u) = \exp(-u)$. Substituting

$$\int_0^\infty \sin(u\sqrt{t})e^{-u}\,du = \frac{1}{1+t}, \qquad \int_0^\infty \cos(u\sqrt{t})e^{-u}\,du = \frac{\sqrt{t}}{1+t}$$

into (7.12), we obtain $B_n(t)$ in terms of $k = 1, 2, \ldots$:

$$B_1(t) = (1+t)^{-1}, \qquad B_{2k}(t) = tB_{2k-1}(t), \qquad B_{2k+1}(t) = -B_{2k}(t). \tag{7.19}$$

By induction (Problem 7.4), we have the $(n-1)$th derivative

$$B_n^{(n-1)}(t) = (n-1)!(1+t)^{-n}. \tag{7.20}$$

Substituting this into (7.11), we obtain

$$f(r) = \frac{2\Gamma(n)}{\Gamma^2(n/2)} r^{n-1}(1+r^2)^{-n}, \tag{7.21}$$

which coincides with the p.d.f. of the square root of an $F(n,n)$ variable. The conclusion follows. □

Lemma 7.3

The vector $\mathbf{y}$ in (7.18) is 1-symmetric with c.f. generator $\psi_0(\cdot)$ which is given by

$$\psi_0(t) = B^{-1}\left(\frac{1}{2}, \frac{n-1}{2}\right)\int_0^1 \Omega_n(w^{-1}t^2)w^{-1/2}(1-w)^{(n-3)/2}\,\mathrm{d}w, \quad (7.22)$$

where $\Omega_n(\|\mathbf{t}\|^2)$ is the c.f. of the uniform distribution on the unit sphere in R^n as given in Theorem 3.1.

PROOF Consider the decomposition of an i.i.d. Cauchy vector (7.17). The density of r is given by (7.21), so that

$$E(r^{it}) = \frac{2\Gamma(n)}{\Gamma^2(n/2)}\int_0^\infty \frac{r^{it+n-1}}{(1+r^2)^n}\,\mathrm{d}r.$$

The integral on the right is positive for all t (see Problem 7.21). According to Theorem 7.4, the vector $\mathbf{y}$ is 1-symmetric. Therefore, we can compute the c.f. of $\mathbf{y}$ as in the proof of Theorem 3.1:

$$\begin{aligned} E(\exp(i\mathbf{t}'\mathbf{y})) &= E\{\exp(i\|\mathbf{t}\|_1 u_1/\sqrt{w_1})\} \\ &= E\{E\{\exp(iw_1^{-1/2}\|\mathbf{t}\|_1 u_1)|w_1\}\} \\ &= E\{\Omega_n(w_1^{-1}\|\mathbf{t}\|_1^2)\}. \end{aligned}$$

Since $w_1 \sim \mathrm{Be}(\frac{1}{2}, ((n-1)/2))$, we arrive at (7.22). □

We are now ready for the decomposition theorem.

Theorem 7.5

An $n \times 1$ random vector $\mathbf{x}$ is 1-symmetric with c.f. generator $\psi(\cdot)$ if and only if the following equivalent conditions are satisfied:

(i) $\mathbf{x} \stackrel{d}{=} r\mathbf{y}$, where $\mathbf{y}$ is as given in (7.18) with density $p_0(\cdot)$ given by (7.10) and c.f. generator $\psi_0(\cdot)$ given by (7.22), and $r > 0$ is an independent random variable with c.d.f. $F(\cdot)$.

(ii) $\psi(t) = \int_0^\infty \psi_0(rt)\,\mathrm{d}F(r)$, where ψ_0 is given by (7.22) and $F(\cdot)$ is the c.d.f. of a positive random variable r.

Furthermore, when the c.f. generator $\psi(\cdot)$ satisfies (7.8), the p.d.f. of $F(\cdot)$ is given by (7.11).

PROOF The equivalence of the two conditions is trivial. The sufficiency of (i) follows from Theorem 7.4. When $\psi(\cdot)$ satisfies (7.8), the necessity of (i) follows from the preceding lemmas. To show the necessity of (i) when $\psi(\cdot)$ does not satisfy (7.8), we consider a perturbation of $\mathbf{x}$, namely $\mathbf{x}+m^{-1}\mathbf{z}$, where $\mathbf{z}$ is composed of i.i.d. standard Cauchy variables independent of $\mathbf{x}$. This perturbation is also 1-symmetric with c.f. generator $\psi(u)\exp(-u/m)$ satisfying (7.8), so that it can be decomposed as

$$\mathbf{x}+m^{-1}\mathbf{z} \stackrel{d}{=} r_m \mathbf{y}, \tag{7.23}$$

where $\mathbf{y}$ is the same for all $m \geqslant 1$ and the p.d.f. of r_m is given by (7.11). Since the c.d.f. of the left-hand side converges pointwise to that of $\mathbf{x}$, we obtain the decomposition of $\mathbf{x}$ as $m \to \infty$. The proof is complete. □

The original proof of this theorem by Cambanis *et al.* (1983) involves many interesting and important results which are given in Problems 7.5 to 7.12 for reference. The present modification which allows us to shorten the proof, and Lemma 7.4 below, are due to K.W. Ng.

Following the pattern in the preceding chapters, we shall call $\mathbf{y}$ the **CKS base** (or a **CKS vector** in more general context), r the **generating variate**, and F the **generating c.d.f.** of $\mathbf{x} \sim S_n(1,\psi)$. Note that the theorem establishes a one-to-one relation between $\psi \in \Psi_n(1)$ and c.d.f.s of nonnegative random variables. The following lemma is another representation of the relation.

Lemma 7.4
Let $\mathbf{z}$ be an i.i.d. standard Cauchy vector, $\mathbf{z} \stackrel{d}{=} r_0\mathbf{y}$, where r_0 has p.d.f. (7.21), $\mathbf{y}$ is as in (7.18). A necessary and sufficient condition that $\mathbf{x} \stackrel{d}{=} r\mathbf{y}$, where r and $\mathbf{y}$ are independent, is that $r_0\mathbf{x} \stackrel{d}{=} r\mathbf{z}$, where both sides are independent products.

PROOF If $\mathbf{x} \stackrel{d}{=} r\mathbf{y}$, then $r_0\mathbf{x} \stackrel{d}{=} r_0 r\mathbf{y} = r(r_0\mathbf{y}) \stackrel{d}{=} r\mathbf{z}$, since r, r_0 and $\mathbf{y}$ are independent. Conversely, for any constant $\mathbf{a}$, $r_0\mathbf{a}'\mathbf{x} \stackrel{d}{=} r\mathbf{a}'\mathbf{z} \stackrel{d}{=} r_0 r\mathbf{a}'\mathbf{y}$, so that $\mathbf{a}'\mathbf{x} \stackrel{d}{=} \mathbf{a}'(r\mathbf{y})$ for any $\mathbf{a}$ according to Theorem 7.4 (cf. the proof of Lemma 7.3), hence $\mathbf{x} \stackrel{d}{=} r\mathbf{y}$. □

Note that, when expressed in terms of a c.f. generator, the relation $r_0\mathbf{x} \stackrel{d}{=} r\mathbf{z}$ means

$$\int_0^\infty \psi(us) f(u)\,\mathrm{d}u = \int_0^\infty e^{-rs}\,\mathrm{d}F(r), \qquad s>0, \tag{7.24}$$

where $f(\cdot)$ is given by (7.21) and $F(\cdot)$ is the c.d.f. of r, the generating variate of $\mathbf{x}$. Thus we immediately obtain the following characterization of $\Psi_n(1)$, which was shown by Cambanis *et al.* (1983) using a different approach.

Theorem 7.6
A function ψ from R_+ to R is the c.f. generator of a 1-symmetric distribution if and only if the function

$$\phi(s) = \int_0^\infty \psi(rs) f(r)\,\mathrm{d}r, \tag{7.25}$$

where $f(\cdot)$ is given by (7.21), is a Laplace transform of a nonnegative random variable. In particular, $\psi \in \Psi_n(1)$ if and only if the function (7.25) is completely monotone, i.e. it has derivatives $\phi^{(n)}(s)$ of all order and

$$(-1)^n \phi^{(n)}(s) \geqslant 0, \qquad s > 0.$$

PROOF The proof concerning the complete monotonicity follows immediately from Bernstein's theorem (see, e.g., Feller, 1971, p. 439). □

7.3 Properties of 1-symmetric distributions

The decomposition of 1-symmetric distributions provides a convenient means to examine the mixed moments:

$$E\left(\prod_{i=1}^n x_i^{m_i}\right) = E(r^m) E\left(\prod_{i=1}^n u_i^{m_i}\right) E\left(\prod_{i=1}^n w_i^{-m_i/2}\right), \qquad m = \sum m_i,$$

where the first factor depends on the distribution of the particular generating variate r, and the second factor is given by (3.6). Since $\mathbf{w} \sim D_n(\frac{1}{2}, \ldots, \frac{1}{2})$, the Dirichlet integral involved with the third factor exists if and only if $m_i < 1$ for all $i = 1, \ldots, n$. It follows that the means and higher mixed moments of a 1-symmetric distribution do not exist. As for marginal distributions, it is obvious that if $(x_1, \ldots, x_n) \sim S_n(1, \psi)$ then $(x_1, \ldots, x_m) \sim S_m(1, \psi)$, so that the marginal p.d.f. is again given by (7.9) and the corresponding generating p.d.f. is given by (7.11) with n replaced by m in both formulas. One may anticipate a relationship, analogous to (2.23) and (6.31), between the generating p.d.f.s for $(x_1, \ldots, x_n)$ and $(x_1, \ldots, x_m)$. One approach is to find a simple relationship between $B_n^{(n-1)}(t)$ and $B_m^{(m-1)}(t)$, where $B_n(\cdot)$ is given by (7.12), and this is quite hard. Another approach is obtaining a relation between the generating variates. Recall that a vector $\mathbf{y}$ having the distribution given in Lemma 7.2 is called a CKS vector of dimension n.

Lemma 7.5

Let a CKS vector $\mathbf{y}$ of dimension n be partitioned into m subvectors $\mathbf{y}^{(i)}$, each of dimension $n_i, i = 1, \ldots, m$. Let $\mathbf{y}_1, \mathbf{y}_2, \ldots, \mathbf{y}_m, \mathbf{w}$ and $\mathbf{v}$ be mutually independent, where $\mathbf{y}_i$ is a CKS vector of dimension n_i, $\mathbf{w} \sim D_m(n_1/2, \ldots, n_m/2)$, $\mathbf{v} \sim D_m(n_1/2, \ldots, n_m/2)$. The following stochastic representation of $\mathbf{y}$ is true:

$$\begin{bmatrix} \mathbf{y}^{(1)} \\ \vdots \\ \mathbf{y}^{(m)} \end{bmatrix} \stackrel{d}{=} \begin{bmatrix} \sqrt{v_1/w_1}\,\mathbf{y}_1 \\ \vdots \\ \sqrt{v_m/w_m}\,\mathbf{y}_m \end{bmatrix}. \tag{7.26}$$

PROOF Being a CKS vector, $\mathbf{y}$ can be represented as $(y_1, \ldots, y_n) \stackrel{d}{=} (u_1/|u'_1|, \ldots, u_n/|u'_n|)$, where $(u_1, \ldots, u_n)$ and $(u'_1, \ldots, u'_n)$ are independently uniform on the unit sphere in R^n. The argument in the proof of Theorem 2.6 can be applied to these two uniform vectors separately. Combining these results, we obtain (7.26) (see Section 1.2.2). □

Letting $m = 2$ in the above lemma, the theorem below follows from the decomposition of a 1-symmetric vector $\mathbf{x}$.

Theorem 7.7

Let $\mathbf{x} \sim S_n(1, \psi)$ be partitioned into subvectors $\mathbf{x}^{(1)}$ and $\mathbf{x}^{(2)}$ of dimensions m and $(n - m)$ respectively, $1 \leqslant m < n$. It is true that

$$\begin{bmatrix} \mathbf{x}^{(1)} \\ \mathbf{x}^{(2)} \end{bmatrix} \stackrel{d}{=} r \begin{bmatrix} \sqrt{\dfrac{v}{w}}\,\mathbf{y}_1 \\ \sqrt{\dfrac{1 - v^2}{1 - w}}\,\mathbf{y}_2 \end{bmatrix}, \tag{7.27}$$

where $r, v, w, \mathbf{y}_1$ and $\mathbf{y}_2$ are mutually independent, $\mathbf{y}_1$ and $\mathbf{y}_2$ are CKS vectors of dimensions m and $(n - m)$ respectively, $v \sim \mathrm{Be}(m/2, (n - m)/2)$, $w \sim \mathrm{Be}(m/2, (n - m)/2)$, and r is the generating variate of $\mathbf{x}$. In particular, $\mathbf{x}^{(1)} \sim S_m(1, \psi)$ with CKS base $\mathbf{y}_1$ and generating variate r_m which can be represented as

$$r_m \stackrel{d}{=} r\sqrt{v/w}. \tag{7.28}$$

We need the following lemma to provide an analytic relationship between the densities of r_m and r.

Lemma 7.6

The p.d.f. of $x = \sqrt{v/w}$, where v and w are i.i.d. Be$(m/2, (n-m)/2)$, is given by

$$q(x) = \begin{cases} x^{m-1}h(x^2), & 0 < x < 1, \\ x^{-m-1}h(1/x^2), & x > 1, \end{cases} \tag{7.29}$$

where the function $h(\cdot)$ is defined as

$$h(t) = \frac{2\Gamma(m)\Gamma^2(n/2)}{\Gamma^2(m/2)\Gamma((n-m)/2)\Gamma((n+m)/2)} \sum_{k=0}^{\infty} \frac{m^{[k]}(1-(n-m)/2)^{[k]}}{((n+m)/2)} \frac{t^k}{k!}, \tag{7.30}$$

which is uniformly convergent in $-1 < t < 1$ and is absolutely convergent at $t = 1$ when $n - m > 1$.

PROOF Since v and w are identically distributed, x and x^{-1} have identical p.d.f.s. Thus we need only consider $0 < x < 1$. Given a fixed value w, the conditional p.d.f. of $y = v/w$ is easy to get:

$$p(y|w) = B^{-1}(m/2, (n-m)/2)y^{m/2-1}w^{m/2}(1-yw)^{(n-m)/2-1}, \quad 0 < y < w^{-1}.$$

Using a binomial series expansion for $0 < y < 1$ and $0 < w < 1$, we have

$$p(y|w) = \frac{y^{m/2-1}w^{m/2}}{B(m/2,(n-m)/2)} \sum_{k=0}^{\infty} \left(1 - \frac{n-m}{2}\right)^{[k]} \frac{w^k y^k}{k!}.$$

Therefore the unconditional p.d.f. of y is

$$\begin{aligned} p(y) &= \int_0^1 p(y|w)B^{-1}(m/2,(n-m)/2)w^{m/2-1}(1-w)^{(n-m)/2-1}\,dw \\ &= B^{-2}(m/2,(n-m)/2)y^{m/2-1}\sum_{k=0}^{\infty}(1-(n-m)/2)^{[k]} \\ &\quad \times B(m+k,(n-m)/2). \end{aligned}$$

Noting that

$$\Gamma(m+k)=\Gamma(m)m^{[k]}, \qquad \Gamma((n+m)/2+k)=\Gamma((n+m)/2)((n+m)/2)^{[k]},$$

we obtain

$$p(y)=\frac{\Gamma(m)\Gamma^2(n/2)y^{m/2-1}}{\Gamma^2(m/2)\Gamma((n-m)/2)\Gamma((n+m)/2)}\sum_{k=0}^{\infty}\frac{m^{[k]}(1-(n-m)/2)^{[k]}}{((n+m)/2)^{[k]}}\frac{y^k}{k!}.$$

Now the p.d.f. of x is derived by a further transformation $y=x^2$. The convergence properties of the power series in (7.30) are elementary results in mathematical analysis. □

Corollary
The generating density f of $(x_1,\ldots,x_n)\sim S_n(1,\psi)$ and the generating density f_m of $(x_1,\ldots,x_m)$, $m<n$, are related as

$$f_m(x)=\int_0^{\infty} q(r^{-1}x)f(r)\,\mathrm{d}r, \tag{7.31}$$

where $q(\cdot)$ is given by (7.29).

The independence structure displayed in Theorem 7.7 allows easy identification of the conditional distribution for a subvector. Since $\mathbf{y}_1$ is independent of r, v, w and $\mathbf{y}_2$, any conditioning using the last four variables does not affect $\mathbf{y}_1$. Thus we have the following theorem.

Theorem 7.8
Let an $n\times 1$ 1-symmetric vector $\mathbf{x}$ be partitioned as in Theorem 7.7. For almost every value of $\mathbf{x}^{(2)}$, the conditional distribution of $\mathbf{x}^{(1)}$ given $\mathbf{x}^{(2)}$ is determined and has a 1-symmetric distribution with its generating variate being represented as

$$r_{1\cdot 2}\stackrel{d}{=}(r\sqrt{v/w}\,|\,r\sqrt{(1-v^2)/(1-w^2)}\mathbf{y}_2=\mathbf{x}^{(2)}),$$

where r, v, w and $\mathbf{y}_2$ are as given in Theorem 7.7.

We have seen in Chapter 2 that the orthogonal matrices form the maximal group of linear transformations under which the spherical (2-symmetric) distributions are invariant. For the 1-symmetric distributions here it turns out that the maximal group of matrices is generated by permuting the rows and/or columns of diagonal matrices with nonzero elements being ± 1.

Theorem 7.9

Let $\mathbf{x} \sim S_n(1, \psi)$ and $\mathbf{A}$ an $n \times n$ matrix. Then $\mathbf{Ax} \overset{d}{=} \mathbf{x}$ if and only if on each row and column of $\mathbf{A}$ there is one and only one nonzero element which is either 1 or -1.

PROOF Note that in general $\mathbf{Ax} \overset{d}{=} \mathbf{x}$ if and only if $\mathbf{c}'\mathbf{Ax} \overset{d}{=} \mathbf{c}'\mathbf{x}$ for all $\mathbf{c} \neq \mathbf{0}$. Since $\mathbf{x} \sim S_n(\alpha, \psi)$, this is equivalent to $\|\mathbf{A}'\mathbf{c}\|_1 x_1 \overset{d}{=} \|\mathbf{c}\|_1 x_1$ for all $\mathbf{c} \neq \mathbf{0}$, which in turn is equivalent to

$$\|\mathbf{A}'\mathbf{c}\|_1 = \|\mathbf{c}\|_1, \quad \text{for all} \quad \mathbf{c} \neq \mathbf{0}. \tag{7.32}$$

Using special values of $\mathbf{c}$, we can derive properties about the elements of $\mathbf{A}$. Letting $\mathbf{c}' = (1, 0, \ldots, 0)$, we see that the first column $\mathbf{a}_1$ of $\mathbf{A}'$ satisfies $\|\mathbf{a}_1\|_1 = 1$. Similarly, we see that the jth column $\mathbf{a}_j$ of $\mathbf{A}'$ satisfies

$$\|\mathbf{a}_j\|_1 = 1, \qquad j = 1, \ldots, n. \tag{7.33}$$

Letting $\mathbf{c}' = (1, 1, 0, \ldots, 0)$, we get

$$\sum_{i=1}^{n} |a_{i1} + a_{i2}| = \sum_{i=1}^{n} |a_{i1}| + \sum_{i=1}^{n} |a_{i2}|, \tag{7.34}$$

which is true if and only if a_{i1} and a_{i2} have same signs, $i = 1, \ldots, n$. Now setting $\mathbf{c}' = (1, -1, 0, \ldots, 0)$, we get

$$\sum_{i=1}^{n} |a_{i1} - a_{i2}| = \sum_{i=1}^{n} |a_{i1}| + \sum_{i=1}^{n} |a_{i2}|, \tag{7.35}$$

which is true if and only if a_{i1} and a_{i2} have opposite signs, $i = 1, \ldots, n$. Therefore, a_{i1} and a_{i2} cannot be both nonzero, $i = 1, \ldots, n$. Using this argument for all pairs of columns of $\mathbf{A}'$, we see that there is at most one nonzero element on each row and each column of $\mathbf{A}$. Now the condition (7.33) implies that on each column of $\mathbf{A}'$, there is one nonzero element which is either 1 or -1. The proof is therefore complete. □

Applying Theorem 7.2 and Theorem 7.3, we have two more properties of 1-symmetric distributions.

Theorem 7.10

The components of a 1-symmetric vector $\mathbf{x}$ are i.i.d. scaled Cauchy if and only if $\mathbf{x}$ contains two independent subvectors.

Theorem 7.11
Let $\mathbf{t}(\mathbf{x})$ be a k-dimensional statistic of a 1-symmetric vector of dimension n. If $\mathbf{t}(c\mathbf{x}) \stackrel{d}{=} \mathbf{t}(\mathbf{x})$ for all $c > 0$, then $\mathbf{t}(\mathbf{x}) \stackrel{d}{=} \mathbf{t}(\mathbf{y}) \stackrel{d}{=} \mathbf{t}(\mathbf{z})$, where $\mathbf{y}$ is a CKS vector and $\mathbf{z}$ consists of i.i.d. standard Cauchy components.

Lemma 7.6 and Theorems 7.9 and 7.11 are due to K.W. Ng.

7.4 Some special cases of $\Psi_n(\alpha)$

Witnessing the scantiness of results concerning the α-symmetric distributions for the cases other than $\alpha = 1, 2$, one is tempted to conclude that this area of distribution theory is in its early stages. In fact, the very existence of the class $\Psi_n(\alpha)$ for $\alpha > 2$ and general n is still an open problem. For $0 < \alpha \leqslant 2, \Psi_n(\alpha)$ contains at least the symmetric stable laws. But since the function $\exp(-\lambda|t|^{\alpha})$ is no longer a characteristic function for any probability distribution for $\alpha > 2$ (see Zolotarev, 1985, p. 2, for the history of this fact), the nonemptiness of the class $\Psi_n(\alpha)$ is not guaranteed. Note that $\Psi_n(\alpha)$ depends on both n and α and there is an abundance of combinations for consideration. We summarize below some facts known to the authors of this volume, without providing detailed proofs.

1. $0 < \alpha \leqslant 2, \Psi_n(\alpha)$ nonempty for any $n \geqslant 1$.
2. When $\alpha = 1$ or 2, the decomposition $\mathbf{x} \stackrel{d}{=} r\mathbf{y}$ exists.
3. $\Psi_2(1/2)$ is conjectured to be decomposable by Cambanis, Keener and Simons (1983), based on the following result. Let $(y_1, y_2) = (u_1/w_1\sqrt{w_2}, u_2/(1-w_1)\sqrt{1-w_2})$, where $\mathbf{u}$ is uniform on S_2 and $w_i \sim \mathrm{Be}(\frac{1}{2}, \frac{1}{2}), i = 1, 2$, and $\mathbf{u}, w_1$ and w_2 are independent. It can be shown that $\mathbf{y}$ is 1/2-symmetric (see Problem 7.16). They also have a proof (but not published, it seems) that every $\psi \in \Psi_2(1/2)$ is a scale mixture in a general sense, i.e. the mixing measure for some members of $\Psi_2(1/2)$ is a signed measure on $[0, \infty)$.
4. The nonemptiness of $\Psi_2(\alpha)$ for $\alpha > 2$ is established by Kuritsyn and Shestakov (1984). They show that the function

$$\psi(u) = \exp(-u), \tag{7.36}$$

which has been demonstrated to be a member of $\Psi_2(\alpha)$, $1 \leqslant \alpha \leqslant 2$, by Bretagnolle, Dacunha-Castelle and Krivine (1966), is also a c.f. generator for $\alpha > 2$ and $n = 2$. Their proof is along the general approach of establishing the nonnegativeness of the Fourier transform of $\psi(\|\mathbf{t}\|_\alpha) = \exp(-\sqrt[\alpha]{|t_1|^\alpha + |t_2|^\alpha})$.

5. The class $\Psi_2(\infty)$, which contains the function ψ in (7.36), is nonempty. Any vector $\mathbf{x} \sim S_2(\infty, \psi)$ can be decomposed as $\mathbf{x} \overset{d}{=} r\mathbf{y}$, where $\mathbf{y}$ is a CKS base of dimension 2 and $r > 0$ is an independent variable. This is also shown by Kuritsyn and Shestakov (see Problems 7.17 and 7.18). The restriction $n = 2$ is important here because the following relationship between the ℓ_1-norm and ℓ_∞-norm,

$$\|\mathbf{t}\|_\infty = \max\{|t_1|, |t_2|\} = \frac{|t_1 + t_2| + |t_1 - t_2|}{2} = \|\mathbf{G}\mathbf{t}\|_1, \quad (7.37)$$

where $\mathbf{G}$ is a matrix of orthogonal rows and columns, cannot be extended to $n \geqslant 3$.

6. For the class $\Psi_3(\infty)$, Kuritsyn and Shestakov (1984) have found some grounds to conjecture that it is empty. They have shown that the plausible function $\psi(\cdot)$ in (7.36) is no longer a c.f. generator. In fact, they have established the following: if $\psi(\cdot)$ is a function such that for some sequence of numbers a_k,

$$\psi^k(u/a_k) \to \exp(-\lambda u^\alpha) \quad \text{as} \quad k \to \infty, \quad (7.38)$$

where $\lambda > 0$, $1 \leqslant \alpha \leqslant 2$, then ψ is not a member of $\Psi_3(\infty)$. Note that if $\Psi_3(\infty)$ is empty, so is $\Psi_n(\infty)$ for all $n \geqslant 3$, because of the fact that $\Psi_n(\alpha) \supset \Psi_{n+1}(\alpha)$.

7. The class $\Psi_\infty(\alpha) = \bigcap_{n=1}^\infty \Psi_n(\alpha)$ has been studied by Bretagnolle *et al.* (1966). It turns out that $\Psi_\infty(\alpha)$ is nonempty for $0 < \alpha \leqslant 2$ and that for any infinite-dimensional vector $\mathbf{x} \sim S_\infty(\alpha, \psi)$ there is a decomposition $\mathbf{x} \overset{d}{=} r\mathbf{y}$, where $\mathbf{y}$ consists of i.i.d. components having a symmetric stable distribution of index α, i.e.

$$\psi(u) = \exp(-u^\alpha). \quad (7.39)$$

A proof of this can be obtained by slightly modifying the proof of Theorem 2.21, with the ℓ_2-norm replaced by the ℓ_α-norm. See Problem 7.20.

Problems

7.1 Verify Corollary 1 to Theorem 7.1.

7.2 Prove Theorem 7.3.

7.3 Let $h(\cdot)$ be a continuous function and $g(\cdot)$ have the $(n-1)$th derivative. Show by induction that

$$\int_0^\infty \cdots \int_0^\infty h\left(\sum_{i=1}^n t_i\right) g^{(n-1)}\left(\sum_{i=1}^n t_i x_i\right) \mathrm{d}t_1 \ldots \mathrm{d}t_n$$
$$= \int_0^\infty h(u) \sum_{k=1}^n \frac{g(ux_k)}{\prod_{j\neq k}(x_k - x_j)} \mathrm{d}u,$$

provided that the integral exists.

7.4 Let $B_n(t)$ be defined iteratively:

$$B_1(t) = (1+t)^{-1}, \quad B_{2k}(t) = tB_{2k-1}(t), \quad B_{2k+1}(t) = -B_{2k}, \quad k = 1, 2, \ldots.$$

Verify by induction that the $(n-1)$th derivative is given by

$$B_n^{(n-1)}(t) = (n-1)!(1+t)^{-n}.$$

7.5 Show that if $w \sim \mathrm{Be}(\frac{1}{2}, \frac{1}{2})$, then for all real numbers s and t,

$$\frac{s^2}{w} + \frac{t^2}{1-w} \overset{d}{=} \frac{(|s|+|t|)^2}{w}.$$

Hint: Transform $w = \sin^2\theta$ and express the left-hand side as $(s+t)^2(1+T_x^2(\theta))$, where $T_x(\theta) = (x + \cos 2\theta)/\sin 2\theta$ and $x = (s-t)/(s+t)$. Then show that for all x, $-1 \leqslant x \leqslant 1$,

$$P(|T_x(\theta)| \leqslant z) = \frac{2}{\pi}\sin^{-1} z, \qquad z > 0.$$

7.6 Using the recursive property of the Dirichlet distribution $D_n(\alpha_1, \ldots, \alpha_n)$ discussed at the end of Section 6.2, show that

$$(w_1, \ldots, w_n) \overset{d}{=} ((1-w_n)w_1', \ldots, (1-w_n)w_{n-1}', w_n),$$

where $(w_1, \ldots, w_n) \sim D_n(\frac{1}{2}, \ldots, \frac{1}{2})$, and $(w_1', \ldots, w_{n-1}') \sim D_{n-1}(\frac{1}{2}, \ldots, \frac{1}{2})$ independently of $(w_1, \ldots, w_n)$.

7.7 Show that for $n \geqslant 3$ and real numbers $t_1, \ldots, t_n$,

$$\frac{t_1}{w_1} + \cdots + \frac{t_n}{w_n} \overset{d}{=} \frac{1}{1-w_n}\left(\frac{t_1}{w_1'} + \cdots + \frac{t_{n-1}}{w_{n-1}'}\right) + \frac{t_n^2}{w_n},$$

where w_i and w_i' are as given in the preceding problem.

7.8 Using Problem 7.5 as a starting point, show by induction that

$$\frac{t_1^2}{w_1}+\cdots+\frac{t_n^2}{w_n}\stackrel{d}{=}\frac{(|t_1|+\cdots+|t_n|)^2}{w_1},\qquad n\geqslant 2,\quad t_i \text{ real numbers.}$$

7.9 Verify Lemma 7.3 by means of the result in the preceding problem.

7.10 Let (u_1,u_2), $u_1^2+u_2^2=1$, be uniformly distributed on the unit circle in R^2, and w be independently $\mathrm{Be}(\frac{1}{2},\frac{1}{2})$. Show that the joint p.d.f. of (x,y), where $x=u_1/w^{1/2}$ and $y=u_1/(1-w)^{1/2}$, is given by

$$p(x,y)=\frac{1}{\pi^2|x^2-y^2|},\qquad |x|<1\leqslant|y| \quad\text{or}\quad |y|<1\leqslant|x|.$$

Hint: Evaluate the c.d.f. $P(u_1/\sqrt{w_1}<x, u_2/\sqrt{1-w}<y)$ with the transformation $s=u_1/\sqrt{w}, t=u_2/\sqrt{1-w}$, which has Jacobian

$$w^{3/2}(1-w)^{3/2}(1-u_1^2)^{1/2}|w-u_1^2|^{-1}.$$

7.11 Let $(u_1,\ldots,u_n)$ have a uniform distribution on the surface of the unit sphere in R^n and let $(u_1',\ldots,u_{n-1}')$ be independently uniform on the surface of the unit sphere in R^{n-1}. Verify that

$$(u_1,\ldots,u_n)\stackrel{d}{=}(u_1'\sqrt{1-u_n}, u_2'\sqrt{1-u_n},\ldots,u_{n-1}'\sqrt{1-u_n},u_n).$$

Hint: Make use of the decomposition $\mathbf{x}=r\mathbf{u}$ for $\mathbf{x}\sim N(0,I)$.

7.12 Using Problems 7.10 and 7.11, verify Lemma 7.2 as follows. Let $(u_1',\ldots,u_{n-1}')$ be uniform on the surface of unit sphere in R^{n-1} and $(w_1',\ldots,w_{n-1}')$ be independently $D_{n-1}(\frac{1}{2},\ldots,\frac{1}{2})$ and suppose that $y_i'=u_i'/\sqrt{w_i'}$, $i=1,\ldots,n-1$, have joint p.d.f. given by (7.10) with n replaced by $(n-1)$. Show that the vector base in Lemma 7.2 can be represented as

$$(y_1,\ldots,y_n)\stackrel{d}{=}\left(y_1'\sqrt{\frac{1-u_n^2}{1-w_n}},\ldots,y_{n-1}'\sqrt{\frac{1-u_n^2}{1-w_n}},\frac{u_n}{w_n}\right).$$

Making the transformation $(y_1',\ldots,y_{n-1}',u_n,w_n)\to(y_1,\ldots,y_n,w_n)$, where

$$y_n=\frac{u_n}{\sqrt{w_n}},\quad y_i=y_i'\sqrt{\frac{1-u_i^2}{1-w_i}},\qquad i=1,2,\ldots,n-1,$$

which has Jacobian $w_n^{-1/2}\{(1-u_n^2)/(1-w_n)\}^{(n-1)/2}$, noting that $w_n \sim \text{Be}(1/2,(n-1)/2)$ and that u_n has density (3.9) with $k=1$, show that the p.d.f. of $(y_1,\ldots,y_n)$ is again given by (7.10).

7.13 Show that for $n=2$, the c.f. generator (7.22) can be expressed as

$$\psi_0(t)=\frac{2}{\pi}\int_t^\infty \frac{\sin x}{x}\,\mathrm{d}x, \qquad t \geqslant 0.$$

7.14 Verify that the common marginal density $p(\cdot)$ of x and y in Problem 7.10 is given by

$$p(x)=\frac{1}{\pi^2|x|}\ln\left|\frac{1+|x|}{1-|x|}\right|, \qquad x\neq 0.$$

7.15 Derive the following identity of integrals of Cambanis *et al.* (1983):

$$\int_0^{\pi/2} g\left(\frac{s^2}{\sin^2\theta}+\frac{t^2}{\cos^2\theta}\right)\mathrm{d}\theta=\int_0^{\pi/2} g\left(\frac{(s+t)^2}{\sin^2\theta}\right)\mathrm{d}\theta,$$

where $s,t\geqslant 0$ and g is a function such that the integral on either side exists and is finite.

7.16 Let (u_1,u_2) be uniformly distributed on the unit circle in R^2, $w_i \sim \text{Be}(\frac{1}{2},\frac{1}{2})$, $i=1,2$, and $\mathbf{u}$, w_1 and w_2 be independent. Show that the distribution of $(u_1/w_1\sqrt{w_2},u_2/(1-w_1)\sqrt{1-w_2})$ has a c.f. of the form $\psi((\sqrt{t_1}+\sqrt{t_2})^2)$ for some function $\psi(\cdot)$ from R_+ to R.
Hint: Problem 7.5.

7.17 Let x_i be i.i.d. scaled Cauchy with c.f. $\exp(-|t|/2)$. Show that (x_1+x_2,x_1-x_2) has c.f. $\exp\{-\max(|t_1|,|t_2|)\}$.
Hint: Use (7.37).

7.18 Show that if $\mathbf{x}\sim S_2(\infty,\psi)$, then $\mathbf{x}\stackrel{d}{=}r\mathbf{y}$ where $\mathbf{y}$ is a CKS vector of dimension 2 and is independent of $r>0$.
Hint: Use (7.37).

7.19 Verify the following solutions of complex functional equations:

(a) If $g(\cdot)$ is a continuous complex function of a nonnegative variable satisfying the Cauchy functional equation (see, e.g. Galambos and Kotz, 1978)

$$g(x+y)=g(x)g(y), \qquad x>0,\ y>0,$$

then $g(x)=\exp(cx)$, where c is a complex number.

(b) If h is a continuous complex function of a nonnegative variable satisfying the functional equation

$$h((|x|^\alpha + |y|^\alpha)^{1/\alpha}) = h(|x|)h(|y|)$$

for real numbers x, y, then $h(u) = \exp(cu^\alpha)$, where c is a complex number.

7.20 Let x_n, $n = 1, 2, \ldots$, be an infinite sequence of random variables such that for all n, $(x_1, \ldots, x_n) \sim S_n(\alpha, \psi)$ for some fixed $\psi(\cdot)$. Show that

$$\psi(u) = \exp(-u^\alpha),$$

by mimicking the proof of Theorem 2.21 and using the solution of the functional equation given in Problem 7.19(b).

7.21 It is well known that the conjugate Fourier transform

$$h(t) = \int_{-\infty}^{\infty} e^{itz} g^n(z) \mathrm{d}z$$

equals the n-fold convolution of the conjugate Fourier transform

$$\hat{g}(t) = \int_{-\infty}^{\infty} e^{itz} g(z) \mathrm{d}z.$$

Using the fact (Feller, 1971, p. 503) that

$$\int_{-\infty}^{\infty} \frac{e^{itz}}{\cosh z} \mathrm{d}z = \frac{\pi}{\cosh(\pi t/2)},$$

where cosh is the hyperbolic cosine, show that for all real t,

$$\int_0^{\infty} \frac{r^{it+n-1}}{(1+r^2)^n} \mathrm{d}r > 0.$$

Hint: Transform $r = e^z$ and note that an n-fold convolution of a positive function is a positive function.

References

Ahmad, R. (1972) Extension of the normal family to spherical families. *Trabajos Estadist.*, **23**, 51–60.

Aitchison, J. (1986) *The Statistical Analysis of Compositional Data.* London/New York: Chapman and Hall.

Aitchison, J. and Brown, J.A.C. (1957) *The Lognormal distributions: With Special Reference to its Uses in Economics.* Cambridge, England: University Press.

Ali, M.M. (1980) Characterization of the normal distribution among the continuous symmetric spherical class. *J. Roy. Statist. Soc.*, B, **42**, 162–4.

Anderson, T.W. (1984) *An Introduction to Multivariate Statistical Analysis*, 2nd edition. New York: Wiley.

Anderson, T.W., and Fang, K.T. (1982) Distributions of quadratic forms and Cochran's Theorem for elliptically contoured distributions and their applications. *Technical Report No. 53, ONR Contract N00014-75-C-0442*, Department of Statistics, Stanford University, California.

Anderson, T.W., and Fang, K.T. (1987) Cochran's theorem for elliptically contoured distributions. *Sankhya*, **49**, Ser A, 305–15.

Andrews, D.F., and Mallows, C.L. (1974) Scale mixtures of normal distributions, *J. Roy. Statist. Soc.*, B, **36**, 99–102.

Arnold, S.F., and Lynch, J. (1982) On Ali's characterization of the spherical normal distribution. *J. Roy. Statist. Soc.*, B, **44**, 49–51.

Azlarov, T.A., and Volodin, N.A. (1986) *Characterization Problems Associated with the Exponential Distribution.* New York: Springer-Verlag.

Bartlett, M.S. (1934) The vector representation of a sample. *Proc. Camb. Phil. Soc.*, **30**, 327–40.

Bennett, B.M. (1961) On a certain multivariate non-normal distribution. *Proc. Camb. Phil, Soc.*, **57**, 434–6.

Bentler, P.M., and Berkane, M. (1986) Greatest lower bound to the elliptical theory kurtosis parameter. *Biometrika*, **73**, 240–1.

Bentler, P.M., Fang, K.T., and Wu, S.T. (1988) The logelliptical distributions, in preparation.

Beran, R. (1979) Testing for elliptical symmetry of a multivariate density. *Ann. Statist.*, **7**, 150–62.

Berk, R.H. (1986) Sphericity and the normal law. *Ann. of Probab.*, **14**(2), 696–701.

Berkane, M., and Bentler., P.M. (1986) Moments of elliptically distribted random variates. *Statist. and prob. Letters*, **4**, 333–5.

Berkane, M., and Bentler, P.M. (1987) Mardia's coefficient of kurtosis in elliptical populations, to be submitted.
Bishop, L., Fraser, D.A.S., and Ng, K.W. (1979) Some decompositions of spherical distributions. *Statist. Hefte,* **20**, 1–20.
Blake, I.F., and Thomas, J.B. (1968) On a class of processes arising in linear estimation theory. *IEEE. Trans. Inf. Theory,* **14**, 12–16.
Block, H.W. (1985) Multivariate exponential distribution. *Encyclopedia of Statistical Sciences,* vol. 6 (eds S. Kotz, N.L. Johnson, and C.B. Read), New York: Wiley, pp. 55–9.
Box, G.E.P. (1953) Spherical distributions (Abstract). *Ann. Math. Statist.,* **24**, 687–8.
Bretagnolle, J., Dacunha-Castelle, D., and Krivine, J.L. (1966) Lois stables et espaces L^p. *Ann. Inst. Poincaré,* II, **3**, 231–59.
Brown, T.C., Cartwright, D.I., and Eagleson, G.K. (1985) Characterizations of invariant distributions. *Math Proc. Camb. Phil. Soc.,* **97**, 349–55.
Brown, T.C., Cartwright, D.I., and Eagleson, G.K. (1986) Correlations and characterizations of the uniform distribution. *Austral. J. satist.,* **28**, 89–96.
Brunner, L.J. (1988) Bayes procedures for elliptically symmetric distributions. *Technical Report No. 88–5,* Department of Statistics, State University of New York at Buffalo, New York.
Cacoullos, T., and Koutras, M. (1984) Quadratic forms in spherical random variables: generalized noncentral χ^2 distribution. *Naval Res. Logist. Quart.,* **31**, 447–61.
Cambanis, S., Huang, S., and Simons, G. (1981) On the theory of elliptically contoured distributions. *J. Mult. Anal.,* **11**, 368–85.
Cambanis, S., Keener, R., and Simons, G. (1983) On α-symmetric distributions. *J. Multivariate Anal.,* **13**, 213–33.
Cambanis, S., and Simons, G. (1982) Probability and expectation inequalities. *Z. Wahr. verw. Gebiete,* **59**, 1–25.
Cheng, P. (1987) The limit distributions of empirical distributions of random projection pursuit. *Chinese J. Appl. Prob. Statist.,* **3**, 8–20.
Chmielewski, M.A. (1980) Invariant scale matrix hypothesis tests under elliptical symmetry. *J. Mult. Anal.,* **10,** 343–50.
Chmielewski, M.A. (1981) Elliptically symmetric distributions: A review and bibliography. *Inter. Statist. Review,* **49**, 67–74.
Chu, K.C. (1973) Estimation and decision for linear systems with elliptical random processes. *IEEE Transactions On Automation Control,* **AC18**, 499–505.
Clayton, D.G. (1978) A model for association in bivariate life-tables. *Biometrika,* **65**, 141–51.
Cléroux, R., and Ducharme, G.R. (1989) Vector correlation for elliptical distributions. *Commun. Statist.–Theory Meth.,* **18**(4), 1441–54.
Cook, R.D., and Johnson, M.E. (1981) A family of distributions. *J.R. Statist Soc.,* B, **43**, 210–18.
Cox, D.R. (1972) Regression models and life-tables (with discussion). *J.R. Statist. Soc.,* B, **34**, 187–220.
Das Gupta, S., Eaton, M.L., Olkin, I., Perlman, M., Savage, L.J., and Sobel, M. (1972) Inequalities on the probability content of convex regions for

elliptically contoured distributions. *Proc. 6th Berk. Symp.*, **2**, 241–65.
David, H.A. (1981) *Order Statistics*, 2nd edition. New York: John Wiley.
Dempster, A.P. (1969) *Elements of Continuous Multivariate Analysis*. Reading, Mass.: Addison-Wesley.
Deng, W. (1984) Testing for ellipsoidal symmetry, Contributed Papers. *China-Japan Symposium on Statistics*, 55–8.
Deny, J. (1959) Sur l'équation de convolution $\mu = \mu * \sigma$. In *Sem. Theor. Patent. M. Brelot, Fac. Sci. Paris, 1959–1960, 4 ann.*
Devlin, S.J., Gnanadesikan, R., and Kettenring, J.R. (1976) Some multivariate applications of elliptical distributions, in *Essays in Probability and Statistics* (eds S.I. Keda *et al.*), Chapter 24, Tokyo: Shinko Tsusho Co. pp. 365–93.
Diaconis, P. (1987) Theory of Poincaré and de Finetti, Abstract 199–8. *IMS Bulletin*, **10**(2), 63–4.
Diaconis, P., and Freedman, D. (1986) A dozen de-Finetti-style results in search of a theory. *Techn. Report #85*, Department of Statistics, University of California, Berkeley, California.
Dickey, J.M., and Chen, C.H. (1985) Direct subjective-probability modelling using ellipsoidal distributions. *Bayesian Statistics*, **2**, (eds J.M. Bernardo *et al.*), Elsevier Science Publishers and Valencia Univ. Press, pp. 157–82.
Dickey, J.M., Dawid, A.P., and Kadane, J.B. (1986) Subjective-probability assessment methods for multivariate-*t* and matrix-*t* models, in *Bayesian Inference and Decision Techniques* (eds P. Goel and A. Zellner). Amsterdam: Elsevier Science Publishers, pp. 177–95.
Dickey, J.M., and Chen, C.H. (1987) Representations of elliptical distributions and quadratic forms. *J. Chinese Statist. Assoc.*, **25**, Special Issue, 33–42.
Dixon, W.J. (1950) Analysis of extreme values. *Ann. Math. statist.*, **21**, 488–506.
Doksum, K. (1974) Tailfree and neutral random probabilities and their posterior distributions. *Ann. Statist.*, **2**, 183–201.
Eaton, M.L. (1981) On the projections of isotropic distributions. *Ann. Statist.*, **9** 391–400.
Eaton, M.L. (1983) *Multivariate Statistics—a Vector Space Approach*. New York: Wiley.
Eaton, M.L. (1985) The Gauss–Markov theorem in multivariate analysis. *Multivariate Analysis—VI* (ed. P. R. Krishnaiah), Amsterdam: Elsevier Science Publishers B.V., 177–201.
Eaton, M.L. (1986) A characterization of spherical distributions. *J. Mult. Anal.*, **20**, 272–6.
Eaton, M.L. (1986) Concentration inequalities for Gauss–Markov estimators, *Technical Report No. 479*, University of Minnesota.
Eaton, M.L. (1987) Invariance with applications in Statistics.
Edwards, J. (1922) *A Treatise on the Integral Calculus*, Vol. 2. New York, Macmillan.
Efron, B. (1969) Student's t-test under symmetry conditions. *J. Am. Statist. Assoc.*, **64**, 1278–302.
Efron, B., and Olshen, R.A. (1978) How broad is the class of normal scale mixtures. *Ann. of Statistics*, , **6**(5), 1159–1164.

Epstein, B. (1960) Tests for the validity of the assumption that the underlying distribution of life in exponential. *Technometrics*, **2**, 83–101.

Esseen, C. (1945) Fourier analysis of distribution functions. A mathematical study of the Laplace–Gaussian law. *Acta Mathematica*, **77**, 1–125.

Fang, B.Q., and Fang, K.T. (1988) Order statistics of multivariate symmetric distributions related to exponential distribution and their applications. *Chinese J. Appl. Prob. Stat.*, **4**, 46–54.

Fang, K.T. (1987) A review: on the theory of elliptically contoured distributions. *Math. Advance Sinica*, **16**, 1–15.

Fang, K.T. and Bentler, P.M. (1989) A largest characterization of special and related distributions, to be submitted.

Fang, K.T., Bentler, P.M., and Chou, C.P. (1988) Additive logistic *elliptical distributions*, in progress.

Fang, K.T., and Chen, H.F. (1984) Relationships among classes of spherical matrix distributions. *Acta Math. Appl. Sinica* (English Ser.), **1**, 139–47.

Fang, K.T., and Fang, B.Q. (1986) A New Family of Multivariate Exponential Distributions. Academia Sinica, Beijing.

Fang, K.T., and Fang, B.Q. (1988a) Families of multivariate exponential distributions with location and scale parameters, *Northeastern Math. J.* **4**(1), 16–28.

Fang, K.T., and Fang, B.Q. (1988b) Some families of multivariate symmetric distributions related to exponential distribution. *J. Mult. Anal.*, **24**,109–122.

Fang, K.T., and Fang, B.Q. (1988c) A class of generalized symmetric Dirichlet distributions, *Acta Math. Appl. Sinica*, **4**.

Fang, K.T., and Fang, B.Q. (1989) A characterization of multivariate l_1-norm symmetric distribution. *Statistics and Probability Letters*, **7** (1989), 297–299.

Fang, K.T., and Xu, J.L. (1986) The direct operations of symmetric and lower-triangular matrices with their applications. *Northeastern Math. J.*, **2**, 4–16.

Fang, K.T., and Xu, J.L. (1987) The Mills' ratio of multivariate normal and spherically symmetric distributions. *Acta Math. Sinica*, **30**, 211–20.

Fang, K.T., and Xu, J.L. (1989) A class of multivariate distributions including the multivariate logistic, *J. Math. Research and Exposition*, **9**, 91–100.

Fang, K.T., and Zhang, Y. (1989) *Generalized Multivariate Analysis*. Science Press and Springer-Verlag, to appear.

Feller, W. (1971) *An Introduction to Probability Theory and Applications*, vol. II, 2nd edition. New York: John Wiley & Sons.

Fichtenholz, G.M. (1967) *Differential-und integralrechnung*, III. Berlin: Veb dentscher Verlag Der Wissenschaften.

Fisher, R.A. (1929) Tests of significance in harmonic analysis. *Proc. Roy. Soc. A*, **125**, 54–9.

Fisher, R.A. (1925) Applications of Student's distribution. *Metron*, **5**, 90–104.

Fortuin, C.M., Kastelyn, P.W., and Ginibre, J. (1971) Correlation inequalities on some partially ordered sets. *Comm. Math. Phys.*, **22**, 89–103.

Fraser, D.A.S., and Ng, K.W. (1980) Multivariate regression analysis with spherical error, in *Multivariate Analysis V* (ed. P.R. Krishnaiah). New York: North-Holland , pp. 369–86.

Freedman, D.A. (1963) On the asymptotic behaviour of Bayes estimates in the discrete case. *Ann. Math. Statist.*, **34**, 1386–403.

Freund, R.J. (1961) A bivariate extension of the exponential distribution. *J. Am. Statist. Assoc.*, **56**, 971–7.

Galambos, J., and Kotz, S. (1978) *Characterization of Probability Distributions.* Lecture Notes in Mathematics, No. 675. Springer-Verlag.

Gallot, S. (1966) A bound for the maximum of a number of random variables. *J. Appl. Prob.*, **3**, 556–8.

Ghosh, M., and Pollack, E. (1975) Some properties of multivariate distributions with pdf's constant on ellipsoids. *Comm. Statist.*, **4**, 1157–60.

Giri, N. (1988) Locally minimax tests in symmetric distributions. *Ann. Inst. Statist. Math.*, **40**, 381–94.

Giri, N., and Das, K. (1988) On a robust test of the extended GMANOVA problem in elliptically symmetric distributions. *Sankhya Ser. A*, **50**, 205–10.

Goldman, J. (1974) Statistical properties of a sum of sinusoids and gaussian noise and its generalization to higher dimensions. *Bell System Tech. J.*, **53**, 557–80.

Goldman, J. (1976) Detection in the presence of spherically symmetric random vectors. *IEEE Trans. Info. Theory*, **22**, 52–9.

Goodman, I.R., and Kotz, S. (1981) Hazard rate based on isoprobability contours. *Statistical Distributions in Scientific Work*, Vol. 5 (eds G.P. Patil *et al.*) Boston/London: D. Reidel, pp. 289–309.

Goodman, N.R. (1963) statistical analysis based on a certain multivariate Gausisian distribution (an introduction). *Ann. Math. Statist.*, **34**, 152–76.

Gordon, Y. (1987) Elliptically contoured distributions. *Probab. Th. Rel. Fields*, **76**, 429–38.

Graham, A. (1986) *Kronecker Products and Matrix Calculus: with Applications*, New York: Halsted Press.

Gualtierotti, A.F. (1974) Some remarks on spherically invariant distributions. *J. Mult. Anal.*, **4**, 347–9.

Guiard, V. (1986) A general formula for the central mixed moments of the multivariate normal distributions. *Math. Oper. und Stat. Ser. Statistics*, **17**, 279–89.

Gumbel, E.J. (1960) Bivariate exponential distributions. *J. Am. Statist. Assoc.*, **55**, 698–707.

Gumbel, E.J. (1961a) Bivariate logistic distribution. *J. Am. Statist. Assoc.*, **56**, 335–49.

Gumbel, E.J. (1961b) Multivariate exponential distributions. *Bulletin of the International Statistical Institute*, **39**(2), 469–75.

Gupta, S.S. (1963) Probability integrals of multivariate normal and multivariate *J. Am. Math. Statist.*, **34**, 792–828.

Gupta, R.D., and Richards, D.St.P. (1987) Multivariate Liouville distributions. *J. Multi. Anal.*, **23**, 233–56.

Guttman, I., and Tiao, G.C. (1965) The inverted Dirichlet distribution with applications. *J. Amer. Statist. Assoc.*, **60**, 793–805.

Haralic, R.M. (1977) Pattern discrimination using ellipsoidally symmetric multivariate density functions. *Pattern Recognition*, **9**, 89–94.

Hardin, C. (1982) On linearity of regression. *Z. Wahrsch.*, **61**, 293–302.

Harris, R. (1966) *Reliability Applications of a Bivariate Exponential Distri-*

bution. University of California Operations Research Center, ORC 66-36.

Hartman, P., and Wintner, A. (1940) On the spherical approach to the normal distribution law. *Am. J. Math.*, **62**, 759–79.

Herz, C. (1955) Bessel functions of matrix argument. *Ann. Math.*, **61**, 474–523.

Higgins, J.J. (1975) A geometrical method of constructing multivariate densities and some related inferential procedures. *Comm. Statist.*, **4**, 955–66.

Hochberg, Y., and Tamhane, A.C. (1987) *Multiple Comparison Procedures*. New York: John Wiley.

Holmquist, B. (1988) Moments and cumulants of the multivariate normal distribution. *Stochastical Analysis and Applications*, **6**, 273–8.

Hougard, P. (1986) A class of multivariate future time distributions. *Biometrika*, **73**, 671–8.

Hsu, P.L. (1940) An algebraic derivation of the distribution of rectangular coordinates. *Proc. Edin. Math. Soc.*, **6**(2), 185–9.

Hsu, P.L. (1954) On characteristic functions which coincide in the neighborhood of zero. *Acta Math. Sinica*, **4**, 21–31. Reprinted in Pao-Lu Hsu, *Collected Works*, 1983, Springer-Verlag.

Iyengar, S., and Tong, Y.L. (1988) Convexity properties of elliptically contoured distributions, to appear in *Sankhyā*, Ser. A.

Jajuga, K. (1987) Elliptically symmetric distributions and their applications to classification and regression, in *Proc. Second International Tampere Conference in Statistics* (eds T. Pukkila and S. Puntanen). Tampere, Finland: Univ. of Tampere, pp. 491–8.

James, A.T. (1964) Distributions of matrix variates and latent roots derived from normal samples. *Ann. Math. Statist.*, **35**, 475–501.

James, I.R., and Mosimann, J.E. (1980) A new characterization of the Dirichlet distribution through neutrality. *Ann. Statist.*, **8**, 183–9.

Jensen, D.R. (1979) Linear models without moments. *Biometrika*, **66**(3), 611–17.

Jensen, D.R. (1985). Multivariate distributions. *Encyclopedia of Statistical Sciences*, 6, (eds S., Kotz, N.L., Johnson, and C.B. Read), Wiley, pp. 43–55.

Jiang, J.M. (1987) Starlike functions and linear functions of a Dirichlet distributed vector to appear in SIAM.

Johnson, M.E. (1987) *Multivariate Statistical Simulation*. New York: Wiley.

Johnson, M.E., and Ramberg, J.S. (1977) Elliptically symmetric distributions: characterizations and random variate generation. *A.S.A. Proc. Statist. Comp. Sect.*, 262–5.

Johnson, N.L., and Kotz, S. (1970) *Distributions in Statistics: Continuous Univariate Distributions-1*. New York: Wiley.

Johnson, N.L., and Kotz, S. (1972) Distributions in Statistics: *Continuous Multivariate Distributions*. New York: Wiley.

Kariya, T., and Eaton, M.L. (1977) Robust tests for spherical symmetry. *Ann. Statist.*, **5**, 206–15.

Karlin, S. (1968) *Total Positivity*. Vol. 1. Stanford Univ. Press, Stanford, CA.

Karlin, S., and Rinott, Y. (1980s) Classes of orderings of measures and related correlation inequalities. I. Multivariate totally positive distributions. *J. Multivariate Anal.*, **10**, 467–98.

Karlin, S., and Rinott, Y. (1980b) Classes of orderings of measures and related correlation inequalities. II. Multivariate reverse rule distributions. *J. Multivariate Anal.*, **10**, 499–516.

Kelker, D. (1970) Distribution theory of spherical distributions and a location-scale parameter. *Sankhya*, A, **32**, 419–30.

Kelker, D. (1971) Infinite divisibility and variance mixtures of the normal distribution. *Ann. Math. Statist.*, **42**(2), 802–8.

Kemperman, J.H.B. (1977) On the FKG-inequality for measures on a partially ordered space. *Indag. Math.*, **39**, 313–31.

Khatri, C.G. (1970) A note on Mitra's paper 'A density-free approach to the matrix variate beta distribution'. *Sankhyā*, A, **32**, 311–18.

Khatri, C.G. (1987) Quadratic forms and null robustness for elliptical distributions. *Proc. Second International Tampere Conference in Statistics* (eds T. Pukkila and S. Puntanen). Tampere, Finland: University of Tampere, pp. 177–203.

Khatri, C.G. (1988) Some asymptotic inferential problems connected with elliptical distributions. *J. Mult. Anal.*, **27**, 319–33.

Khatri, C.G., and Mukerjee, R. (1987) Characterization of normality within the class of elliptical contoured distributions. *Statistics & Probability Letters*, **5**, 187–90.

Kimball, A.W. (1951) On dependent tests of significance in the analysis of variance. *Ann. Math. Statist.*, **22**, 600–2.

King, M.L. (1980) Robust tests for spherical symmetry and their application to least squares regression. *Ann. Statist.*, **8**, 1165–271.

Kingman, J.F.C. (1972) On random sequences with spherical symmetry. *Biometrika*, **59**, 492–4.

Kotz, S. (1975) Multivariate distributions at a cross-road. In *Statistical Distributions in Scientific Work*, **1**. (eds G.P. Patil, S. Kotz and) J.K. Ord, D. Reidel Publ. Co.

Kounias, E.G. (1968) Bounds for the probability of a union, with applications. *Ann. Math. Statist.*, **39**, 2154–8.

Koutras, M. (1986) On the generalized noncentral chi-squared distribution induces by an elliptical gamma law. *Biometrika*, **73**, 528–32.

Kozial, J. A. (1982) A class of invariant procedures for assessing multivariate normality. To be submitted.

Kreyszig, E. (1983) *Advanced Engineering Mathematics*, 5th edn. Wiley, New York.

Krishnaiah, P.R. (1976) Some recent developments on complex multivariate distributions. *J. Multi. Anal.* **6**, 1–30.

Krishnaiah, P.R., and Lin, J.G. (1984) Complex elliptical distributions. *Technical Report No. 84–19*. Center for Multivariate Analysis, University of Pittsburgh.

Kudina, L.S. (1975) On decomposition of radially symmetric distributions. *Theory Prob. Applic.*, **20**, 644–8.

Kumar, A. (1970) Abstract no. 141: A matrix variate Dirichlet distribution. *Sankhyā*, A, **32**, 133.

Kunitsyn, Y.G. (1986) On the least-squares method for elliptically contoured distributions. *Theor. Prob. and Appl.*, **31**, 738–40.

Kuritsyn, Yu. G. and Shestakov, A.V. (1984) On α-symmetric distributions. *Theor. Probab. Appl.* **29**, 804–06.

Kuritsyn, Yu. G. and Shestakov, A.V. (1985) On α-symmetric distributions. *Pacific J. of Math.*, 118–19.

Laurent, A.G. (1973) Remarks on multivariate models based on the notion of structure, in *Multivariate Statistical Inference* (eds D.G. Kabe and R.P. Gupta), New York: North-Holland, pp. 145–68.

Lawless, J.F. (1972) On prediction intervals for samples from the exponential distribution and prediction limits for system survival, *Sankhyā*, Ser B., **34**, 1–14.

Li, G. (1984) Moments of random vector and its quadratic forms, Contributed papers. *China–Japan Symposium on Statistics*, 134–5.

Lindley, D.V., and Singpurwalla, N.D. (1986) Multivariate distributions for the life lengths of components of a system sharing a common environment. *J. Appl. Prob.*, **23**, 418–31.

Lord, R.D. (1974) The use of the Hankel transform in statistics. I. General theory and examples. *Biometrika*, **41**, 44–55.

Lukacs, E. (1956) Characterization of populations by properties of suitable statistics, *Proc. 3rd Berkeley Symp. Math. Stat. and Prob. 2.* Los Angeles and Berkeley: University of California Press.

Malik, H.J., and Abraham, B. (1973) Multivariate logistic distribution. *Ann. Statist.*, **1**, 588–90.

Marcinkiewicz, J. (1938) Sur les fonctions indèpendants III. *Fundamenta Mathematics*, **31**, 66–102.

Mardia, K.V. (1970) Measures of multivariate skewness and kurtosis with applications. *Biometrika*, **57**, 519–30.

Marshall, A.M., and Olkin, I. (1967a) A multivariate exponential distribution. *J. Am. Statist. Assoc.*, **62**, 30–44.

Marshall, A.W., and Olkin, I. (1967b) A generalized bivariate exponential distribution. *J. Appl. Prob.*, **4**, 291–302.

Marshall, A.W., and Olkin, I. (1979) *Inequalities: Theory of Majorization and its Applications.* New York: Academic Press.

Marshall, A.W., and Olkin, I. (1985) Multivariate exponential distribution. *Encyclopedia of Statistical Sciences*, Vol. 6 (eds S. Kotz, N.L. Johnson, and C.B. Read), New York: Wiley, pp. 59–62.

Maxwell, J.C. (1860) Illustration of the dynamical theory of gases. Part I. On the motions and collisions of perfectly elastic bodies. *Phil. Mag.*, **19**, 19–32.

McGraw, D.K., and Wagner, J.F. (1968) Elliptically symmetric distributions. *IEEE Trans. Inf. Theory*, **IT14**, 110–20.

Mihoc, I. (1987) The connection between the n-dimensional Student distribution and the inverted Dirichlet distribution. *Studia Univ. Babes-Bolyai, Mathematica*, **32**(2), 34–9.

Misiewicz, J. (1982) Elliptically contoured measures on R^∞. *Bull. Acad. Polon. Sci. Math.*, **30** (5–6), 283–90.

Misiewicz, J. (1984) Characterization of the elliptically contoured measures on infinite-dimensional Banach spaces. *Prob. and Mathematical Statist.*, **4**, 47–56.

Misiewicz, J. (1985) Infinite divisibility of elliptically contoured measures. *Bull. Acad. Polon. Sci. Math.*, **33**, 1–2.

Mitchell, A.F.S. (1988a) Statistical manifolds of univariate elliptic distributions, manuscript.

Mitchell, A.F.S. (1988b) The information matrix, skewness tensor and α-connections for the general multivariate elliptic distribution, to appear in *Ann. Int. Statist. Math.* (Tokyo).

Mitchell, A.F.S., and Krzanoski, W.J. (1985) The Mahalanobis distance and elliptic distributions. *Biometrika*, **72**, 464–7.

Mitra, S.K. (1970) A density-free approach to the matrix variate beta distribution. *Sankhyā*, A, **32**, 81–8.

Muirhead, R.J. (1980) The effects of elliptical distributions on some standard procedures involving correlation coefficients, in *Multivariate Statistical Analysis* (ed. R.P. Gupta). New York: North-Holland, pp. 143–59.

Muirhead, R.J. (1982) *Aspects of Multivariate Statistical Theory.* New York: Wiley.

Nash, D., and Klamkin, M.S. (1976) A spherical characterization of the normal distribution. *J. Mult. Anal.*, **5**, 156–8.

Ng, K.W. (1988a) On modelling compositional data, to be submitted.

Ng, K.W. (1988b) Some notes on Liouville distributions, to be submitted.

Nimmo-Smith, I. (1979) Linear regression and sphericity. *Biometrika*, **66**(2), 390–2.

Nomakuchi, K., and Sakata, T.(1988a) Characterizations of the forms of covariance matrix of an elliptically contoured distribution. *Sankhyā*, A, **50**, 205–10.

Nomakuchi, K., and Sakata, T. (1988b) Characterization of conditional covariance and unified theory in the problem of ordering random variables. *Ann. Inst. Statist. Math.*, **40**, 93–9.

Oakes, D. (1982) A model for association in bivariate survival data. *J.R. Statist. Soc.*, B, **44**, 414–22.

Olkin, I., and Rubin, H.(1964) Multivariate beta distribution and independence properties of the Wishart distribution. *Ann. Math. Statist.*, **35**, 261–9.

Patel, J.K., Kapadia, C.H., and Owen, D.B. (1976) *Handbook of Statistical Distributions.* New York: Marcel Dekker, Inc.

Perlman, M.D., and Olkin, I. (1980) Unbiasedness of invariant tests for MANOVA and other multivariate problems. *Ann. Statist.*, **8**, 1326–41.

Poincaré, H. (1912) *Calcul des Probabilité.* Paris: Gauthier-Villars.

Raftery, A.E. (1984) A continuous multivariate exponential distribution. *Commun. Statist.–Theory. Math.*, **13**, 947–65.

Rao, C.R. (1973) *Linear Statistical Inference and its Applications*, 2nd edition. New York: John Wiley.

Rao, C.R. (1983) Deny's theorem and its applications to characterizations of probability distributions, in *Festschrift for E.L. Lehmann*, pp. 348–66. Belmont, CA: Wadsworth.

Richards, D. St. P.(1982) Exponential distributions on Abelian semigroups. Mimeo. Series 1387, Institute of Statistics, Univ. of North Corolina.

Richards, D. St. P. (1984) Hyperspherical models, fractional derivatives, and exponential distributions on matrix spaces. *Sankhyā*, A, **46**, 155–65.

Richards, D. St. P. (1986) Positive definite symmetric functions on finite dimensional spaces, I. Applications of the Radon transform. *J. Multivariate Anal.*, **19**, 280–98.

Rooney, P.G (1972) On the ranges of certain fractional integrals. *Canad. J. Math.*, **24**, 1198–216.

Schoenberg, I.J. (1938) Metric spaces and completely monotone functions. *Ann. Math.*, **39**. 811–41.

Sivazlian, B.D. (1981) On a multivariate extension of the gamma and beta distributions, *SIAM J.Appl. Math.*, **41**, 205–9.

Sobel, M., Uppuluri, V.R.R, and Frankowski (1977) Selected Tables in Mathematical Statistics, 4 *Dirichlet Distributions – Type I.* American Mathematical Society. Providence, Rhode Island.

Stadje, W. (1987) A characterization and a variational inequality for the multivariate normal distribution. *J. Australian Math. Soc.*, **43**, 366–74.

Steck, G.P. (1962) Orthant probabilities for the equicorrelated multivariate normal distribution. *Biometrika*, **49**, 433–45.

Sutradhar, B.C. (1986) On the characteristics function of multivariate student *t*-distribution. *Canad. J. Statist.*, **14**, 329–37.

Sutradhar, B.C. (1988) Author's revision. *Canad. J. Statist.*, **16**, 323.

Szablowski, P.J. (1986) Some remarks on two dimensional elliptically contoured measures with second moments. *Demonstratio Mathematica*, **19**(4), 915–29.

Szablowski, P.J. (1987) On the properties of marginal densities and conditional moments of elliptically contoured measures. *Math. Stat. A Prob. Theory Vol. A*, 237–52, Proceedings 6th Pannonian Symp.

Szablowski, P.J. (1987) On the properties of marginal densities and conditional moments of elliptically contoured measures (eds M.L. Puri *et al.*), Vol. A, pp. 237–52. D. Reidel Publishing Company.

Szablowski, P.J. (1988a) Elliptically contoured random variables and their application to the extension of Kalman filter, to appear in *Comput. Math. Applic.*

Szablowski, P.J.(1988b) Expansions of $E(X \mid Y + \varepsilon Z)$ and their applications to the analysis of elliptically contoured measures, to appear in *Comput. Math. Applic.*

Thomas, D.H. (1970) Some contributions to radial probability distributions statistics, and the operational calculi. Ph.D. dissertation, Wayne State University.

Titterington, D.M., Smith, A.F.M., and Mokov, V.E. (1985) *Statistical Analysis of Finite Mixture Distributions.* New York: Wiley.

Tong, Y.L. (1980) *Probabilistic Inequalities in Multivariate Distributions.* New York: Academic Press.

Traat, I. (1989) Moments and cumulants of multivariate elliptical distributions, to appear in *Acta Univ. Tartuensis.*

Tyler, D.E. (1982) Radial estimates and the test for sphericity. *Biometrika*, **69**, 429–36.

Tyler, D.E. (1983) Robustness and efficiency properties of scatter matrices. *Biometrika*, **70**, 411–20.

Wang, L.Y. (1987) Characterization of elliptically symmetric distribution, to be submitted to *J. Stat. & Appl. Prob.*

Watson, G.S. (1984) *Statistics on Spheres.* New York: Wiley-Interscience.

Weyl, H. (1939) *The Classical Groups.* Princeton, N.J.: Princeton Uni. Press.

Weyl, H. (1952) *Symmetry.* Princeton: Princeton University Press.

Wigner, E.P. (1967) *Symmetries and Reflections.* Bloomington: Indiana University Press.

Williamson, R.E. (1956) Multiply monotone functions and their Laplace transforms. *Duke Math. J.*, **23**, 189–207.

Withers, C.S. (1985) The moments of the multivariate normal. *Bull. Austral. Math. Soc.*, **32**, 103–7.

Wolfe, S.J. (1975) On the unimodality of spherically symmetric stable distribution functions. *J. Mult. Anal.*, **5**, 236–42.

Zhang, Y. and Fang, K.T. (1982) *An Introduction to Multivariate Analysis.* Beijing: Science Press.

Zolotarev, V.M. (1981) Integral transformations of distributions and estimates of parameters of multidimensional spherical symmetric stable laws. *Contributions to Probability: Collection Dedicated to Eugene Lukacs* (eds J. Gani and V.K. Rohatgi), pp. 283–305. New York: Academic Press.

Zolotarev, V.M. (1983) *One Dimensional Stable Distributions*, (in Russian), Moscow. Nauka.

Zolotarev, V.M. (1985) *One-dimensional Stable Distributions.* Translations of Mathematical Monographs, vol. 65. Providence, R.I.: American Mathematical Society.

Index